Projects and Demonstrations
in
Astronomy

Projects and Demonstrations in Astronomy

D. Tattersfield
M.A.(Cantab), C.Eng., F.I.Mech.E., F.R.A.S.

Head of the Department of Mechanical and Production Engineering
North Gloucestershire College of Technology

A HALSTED PRESS BOOK

JOHN WILEY & Sons
New York–Toronto

First published in 1979 by Stanley Thornes (Publishers) Ltd.,
Educa House, 32 Malmesbury Road, Kingsditch,
CHELTENHAM GL51 9PL

Published in the U.S.A., Canada
and Latin America by
Halsted Press
a Division of John Wiley & Sons Inc.
New York

Library of Congress Cataloging in
Publication Data
Tattersfield, D.

   Projects and Demonstrations in Astronomy

ISBN 0470–26715–1
L.C. 79–84264

Preface

Astronomy, like other sciences, is essentially both a practical and a theoretical subject. Centuries of patient observation and measurement have produced a wealth of data about the celestial bodies, from which a detailed knowledge of our universe has been deduced. Sometimes theoretical astronomers forecast the possibility of events, or the presence of bodies not hitherto discovered or of bodies showing certain characteristics, and later such events take place and such bodies are discovered.

All this is the more remarkable when we remember that, apart from the Moon, and that only recently, none of the planets, stars, nebulae, galaxies or even the space between them can be touched by humans, and that their distances are, in general, measured in millions, if not millions of millions, of kilometres. The space vehicles sent unmanned to the nearer planets in the last decade have greatly increased our information, not only about the planets themselves but about the conditions existing in the near environment of the planets, including the Earth.

For anyone studying astronomy it is necessary to be practical and observe the heavens with the naked eye, with binoculars or with a telescope. Unfortunately, our skies are often overcast with cloud for much of the time, and on the other hand not everyone who is interested in astronomy has access to a telescope. This can be a frustration to both teacher and student where astronomy is being taught formally as part of a science course, or as a general study in other courses, or, indeed, where an amateur astronomer is trying to further his or her knowledge.

It is with these thoughts that *Projects and Demonstrations in Astronomy* has been devised. It covers a wide range of assignments, all of which can be carried out indoors independent of weather conditions. There are a large number of projects, all of which, with a very few exceptions, can be undertaken on A4 size pieces of paper (297 mm × 210 mm) with the help of no more than a pencil, a scale, a pair of compasses, a protractor and a set square.

In this book are collected together additionally a number of demonstrations which illustrate certain principles found in astronomy and a number of models have been described which students can make, or the teacher can use for demonstrations. The overall criteria for inclusion are that the demonstrations or teaching aids are non-trivial, that they can easily be constructed from materials which are readily available and, most important, that they are effective.

Each assignment has been preceded by enough astronomical theory for the reader to appreciate its objects, to carry it out and to achieve a worthwhile result. All the projects have a graphical approach and it will be seen throughout the book that even complicated astronomical ideas can be appreciated by this approach. Only in a very few of the projects does the mathematical skill required go further than an appreciation of logarithms and an elementary knowledge of trigonometry.

Many of the projects stem from data which have been obtained by professional astronomers since it is clearly not possible for the reader to collect his or her own. Some of the aids and demonstrations have been collected from various sources listed under 'Acknowledgements'. Almost all have been modified from their original form, and appendices have been added at the back of the book to make the items more interesting and informative to the teacher or to the student. Similarly the solutions to most of the projects are given at the back of the book.

This book, then, offers an approach to practical work in astronomy which can be carried out indoors. Taken together, the theory and assignments form a course in astronomy which is self-supporting and flexible. It is hoped that this volume will prove a real contribution to the teaching and learning of astronomy, whether in the home, in schools, colleges or universities.

December 1977   D. Tattersfield

Acknowledgements

Permission to include in this book some projects and demonstrations devised by others has been freely forthcoming, and for this I am most grateful. A list of these generous contributors is given below.

Bernard R. Ambrose, Keswick Hall, College of Education, Norwich (Section 6.2)

Dr. Laura P. Bautz, National Science Foundation, Washington, D.C. (Section 8.12)

Dr. D. S. Birney, Whitin Observatory, Wellesley College, Wellesley, Massachusetts (Section 7.34)

Dr. David Clarke, Department of Astronomy, University of Glasgow (Sections 7.11, 8.6 and 8.20)

Professor Owen Gingerich, Harvard University and the Smithsonian Astrophysical Observatory (Sections 5.24, 5.26 and 7.9)

Philip S. Heelis, Warwick School, Warwick (Section 7.23)

Dr. Maurice J. Kenn, Imperial College of Science and Technology, London (Section 5.14)

Dr. G. S. Mumford, Tufts University, Medford, Massachusetts (Section 8.9)

L. S. T. Symms, Royal Greenwich Observatory, Herstmonceux (Data for Section 7.37)

Alan Ward, St. Mary's College of Education, Cheltenham (Section 5.22)

Data have been extracted from the following publications.

Memoirs of the British Astronomical Association (data of comet orbits)
Circular of the British Astronomical Association (data of Comet Arend-Roland)
Handbooks of the British Astronomical Association (miscellaneous data)
The Astronomical Ephemeris, H.M.S.O.

I am indebted to those individuals and institutions who have allowed the reproduction of some of the photographs in this book. Credit is given under each photograph. I have also enjoyed discussion with my colleague Michael J. Pullin on mathematical aspects, the results of which appear in the Appendices.

It has been a real pleasure to work with Bryan Foster of the Salisbury Astronomical Society who has prepared the diagrams and with the staff of the publishers, Stanley Thornes (Publishers) Ltd.

Finally I wish to record appreciation of the encouragement given to me by my wife Margaret during the preparation of the book and for her help with the typing of the final version.

Contents

Introduction

This book is about the universe. It is reasonable, therefore, that some preliminary ideas should be given of what objects there are in the universe, so that we shall be ready to study each in more detail in the main body of the book.

Thus we begin with our own Sun, which is accompanied by its planets – Mercury, Venus, Earth, Mars, Jupiter, Saturn, Uranus, Neptune and Pluto, and a whole host of small bodies called asteroids (minor planets) which revolve round the Sun, mostly between the orbits of Mars and Jupiter. If we add to these most of the comets, meteors (or shooting stars), meteorites and the natural satellites (or moons) of some of the planets we have the constituents of *the solar system*.

Our Sun is just one star of about 100 000 millions which form a rotating spiral system, on the whole like a flat disc with a bulge at the centre on both sides of the disc, and with arms forming the spiral. This is our *Galaxy*, and since our Sun is not centrally placed, our view along the plane of the disc shows us a band of stars which we call the Milky Way.

Stars of the sky appear to form groups, or constellations, which seem to change only very slowly with time, and which people in days gone by have imagined as figures of men, animals or gods, though the resemblance in many cases is hard to see. Of the constellations in the northern sky the Plough or Great Bear is probably the best known, while in the southern sky the Southern Cross stands out clearly. It is convenient for us to think of groups of stars as it helps us to find our way about the sky and to give an approximate position of a celestial object by saying that it is in this or in that constellation.

Then we have other, relatively near, galaxies, some of them similar to our own, which we know as *the local cluster of galaxies*, while further out in space we see other clusters of galaxies, each galaxy containing about the same number of stars as our own.

Between the stars the space contains hydrogen and helium gas and smaller quantities of heavier elements, carbon, oxygen, sodium, magnesium and atoms of even heavier elements, iron, nickel, chromium and so on.

Simple molecules are present such as water, carbon dioxide, ammonia and hydrogen cyanide, together with more complex molecules made up of carbon, hydrogen, oxygen, nitrogen and sulphur. Areas of interstellar gas are sometimes illuminated by the light from nearby stars, and we refer to these glowing objects as *nebulae*.

More recently we have the quasars and the pulsars. The former are powerful radio transmitters. They are thought to be at a very great distance from us and are moving away from us at speeds approaching that of light. The latter are also radio transmitters, sending out pulses of energy in the radio-frequency band with such regularity that when the first one was discovered, the thought that this might be an attempt of another civilisation on another planet or star to communicate with us was quite exciting. The discovery of other pulsars has shown this to be incorrect. Pulsars are now fairly well established as being rapidly rotating, very dense (neutron) stars within our own Galaxy.

As indicated in the Preface, theoretical astronomers sometimes forecast the existence of astronomical bodies before they are discovered. One such forecast today is that hyper-dense collections of material exist such that any matter which is attracted into them cannot escape. Nor indeed can light radiation come to us from these so-called black holes. The theoretical evidence for the presence of black holes is very strong, but at the time of writing no black holes have been discovered. Indirectly, however, certain characteristics of a star and its surroundings in the constellation of Cygnus at present are leading astronomers to believe that they have possibly located the position of a black hole. Also, perhaps more positively, a very small region in the galaxy M87 appears to be affecting the motion of nearby stars in the manner which one would expect if a large black hole were located there. Although these ideas are extremely fascinating they are clearly outside the terms of reference of this book.

This then, is the universe in which we live. It is the purpose of this book to say more about most of these objects, and to show how these complex stars and systems of stars can be better understood by carrying out graphical projects and by doing simple demonstrations. The assignments are, in the main, easy to carry out, and the time spent in doing them will be rewarding for those who wish to know more about the universe of which we form such a very small part.

1

Useful devices

1.1 NOTE ON THE MEASUREMENT OF AREAS

Some of the projects in this book depend upon the reader being able to measure areas of irregular shapes.

The simplest way to do this is to draw out the projects on squared paper, and then count the squares included within the boundary of the areas to be measured. Alternatively, a thin piece of transparent graph paper can be placed over the area and again the squares counted. This method can be rather tedious, but none of the projects are so extended that patience becomes exhausted, or even severely tired.

If a planimeter is available, the measurement of such areas presents no problems other than those involved in the manipulation of the instrument. Commercially made planimeters are very expensive, but most colleges with students studying technical subjects usually possess one, often in the thermodynamics laboratory.

Details of a planimeter which can be made for almost nothing are given below.

1.2 THE HATCHET PLANIMETER

This consists simply of a piece of stiff wire, or thin rod, bent as shown in Figure 1.1. The dimensions given are suitable, but not critical. The

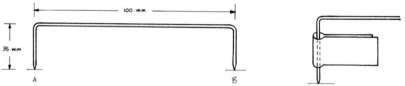

Figure 1.1 The hatchet planimeter

ends of the wire or rod should be pointed, and then the sharp points rounded a little to allow them to slip easily over the surface of the paper.

To use the planimeter, estimate the centre of the area approximately. From this centre O draw a line to any point on the perimeter of the area. Locate one point A of the planimeter on the point O, and make a light impression on the paper with the other point B. Holding the planimeter lightly at the end A, move from O to the perimeter of the area along the line, pass once round the perimeter and then back along the line to O. Make a second light impression with B.

The area is given by multiplying the distance between the two impressions by the distance AB.

Complete freedom of movement of the planimeter is necessary. It will be found beneficial if the vertical part of the planimeter at A is wrapped by a thin strip of paper, and the ends of the paper held by the finger and thumb, thus forming a paper bush for the free movement of the planimeter.

The planimeter, simple though it is, can give accurate results after very little practice. It has even been suggested that a penknife with two blades open to form the shape in Figure 1.1 can be used successfully in this way.

1.3   THE 35 MM SLIDE PROJECTOR

This instrument is one of the most versatile teaching aids for many subjects and especially for astronomy. Apart from projecting 35 mm colour or black and white transparencies on topics of astronomical interest, it can receive within its slide frame a number of devices which assist demonstrations in astronomy. Some of these are described below.

In   the   case   of   a   normal   transparency,   the   film,   which   is

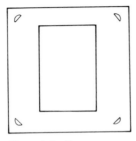

Figure 1.2   Transparency mounting frame

about 35 mm × 38 mm, is sandwiched between the two sides of the frame. In some cases, such as the Agfacolor mounting, four holes in one side of the frame locate as a press fit onto four pegs in the other side of the frame. This is a most suitable type for our purpose (Figure 1.2). However, there are other types, some of which are pre-glued, and the pressure between the two sides is sufficient to retain the film.

(a) Special slide A

In this slide the transparency in the mounting is replaced by a rectangular piece of thin aluminium 35 mm × 38 mm with a hole about 18 mm diameter punched in it centrally (Figure 1.3). It is used with the conic

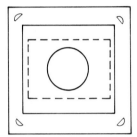

Figure 1.3   Special slide A

section demonstration (Section 2.2) and the artificial sunset demonstration (Section 5.24).

(b) Special slide B

As before, the transparency is replaced by a rectangular piece of thin aluminium 35 mm × 38 mm, but without the large hole. Instead a number of small holes each 0.8 mm diameter (or less if possible) are drilled so as to form a pattern similar to that which is formed by the brighter stars of one of the constellations such as Cassiopeia, Orion, Cygnus and Ursa Major (Figure 1.4).

With such fine holes it is easy to break the slender drill, but this can often be made use of in that the broken, cutting end can then be inserted into the drill with more rigidity. It is advisable before drilling to indent slightly the position of each star hole on the aluminium plate with a very sharp object such as a panel pin. It is also advisable to have a wooden block under the aluminium when the holes are drilled to avoid deformation of the plate. This slide can be used when the major constellations are being described (Section 4.6).

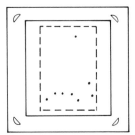

Figure 1.4   Special slide *B*

(c) Special slide C

For this, two thin, double-edged, razor blades are used instead of the transparency or the aluminium plate. They are placed side by side in the place where the transparency would normally fit, thus forming a narrow slit (Figure 1.5). It will be necessary to block the light which might pass

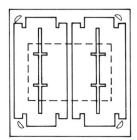

Figure 1.5   Special slide *C*

through the central locating slots in the razor blades. These slots can be covered adequately by opaque strips of gummed paper. This slide is useful when the spectrum demonstration is carried out (Section 7.20).

2

Orbital motion

2.1 THE CONIC SECTIONS

In the past, as in the present, much of the time of astronomers has been taken up in studying the paths which planets, comets and, in some cases, stars follow in relation to their respective parent bodies. For planets and comets the Sun is the parent body. For certain stars the parent body is another star, close enough for the motion to be controlled by the mutual forces of gravity between them. Such stars are referred to as binary stars (see Section 7.1).

The paths which these astronomical bodies follow resemble very closely one or other of the sections revealed when a vertical cone with a circular base is cut by a plane. We refer to the perimeter of these sections as conics. According to the direction of the plane in relation to the sides and axis of the cone, the conic so formed can be a circle, an ellipse, a parabola or a hyperbola.

Conics are also important in astronomy in that the reflecting surfaces of mirrors in astronomical telescopes (see Section 9.1) are mostly parts of conics. Thus that of the main or primary mirror in a reflecting telescope is formed from a parabola, as is the reflecting surface of a radio-telescope of the 'dish' type. The secondary mirror in a Gregorian telescope is formed from an ellipse and that in a Cassegrain telescope from a hyperbola. In a Newtonian reflecting telescope the secondary mirror is plane but the perimeter of this mirror is elliptical since the mirror is set at 45° to the axis of the telescope.

The ways in which the various conic sections are formed by the intersection of a plane and a cone are shown in Figures 2.1–2.4. Truly circular orbits, like truly parabolic orbits, are not common in astronomy, but several of the planets move in nearly circular orbits round the Sun as do some artificial satellites around the Earth. Which sort of conic an astronomical body follows in its orbit depends upon the relation between

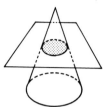

Figure 2.1 Circle –
plane parallel to cone base

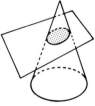

Figure 2.2 Ellipse –
plane not as steep as cone side

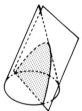

Figure 2.3 Parabola –
plane parallel to cone side

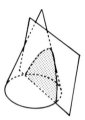

Figure 2.4 Hyperbola –
plane steeper than cone side

its velocity and its distance from the second controlling body at a given time.

If a closer study is made of the motion of the bodies under consideration, it is found that there are variations from the true conics. These are known as *perturbations*. They are caused by the gravitational attractions of astronomical bodies other than the two principal bodies. For instance, the attractions of the planets on the motion of a comet around the Sun will cause perturbations of the orbit of the comet. In most cases these perturbations are too small to be represented graphically in the projects in this book, and for a general study of orbits they will be neglected.

The above method of producing conic sections using a plane and a cone, although essential for the description of these sections, is not very convenient, and further methods of drawing an ellipse, hyperbola and parabola are given in Project 1 (Section 2.4).

2.2 DEMONSTRATION 1

To show the shape of conics using a conical light beam and an intersecting plane.

For this demonstration we shall need a bright cone of light. This can be provided by a slide projector in which is inserted special slide *A* (Section 1.3).

(a) Circle

Place the projector so that the axis of its beam of light is at right angles to a wall. For the purpose of reference call this an east wall. Bring the image into focus, thus obtaining a circle (Figure 2.5(a)).

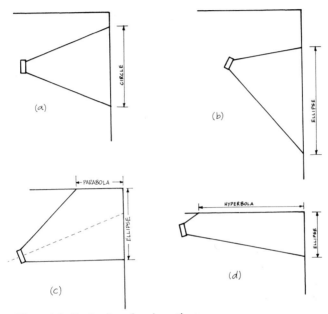

Figure 2.5   Production of conic sections

(b) Ellipse

Carefully, for the lamp filament is at risk when moved hot, turn the projector through an angle of about 45°. The image on the east wall will now be an ellipse (Figure 2.5(b)).

(c) Parabola

Now move the projector close to the north wall with one edge of the beam parallel to the north wall so that the image falls on *both* the north and east walls. That part of the image on the north wall will be a parabola (Figure 2.5(c)) and that on the east wall part of an ellipse.

(d) Hyperbola

Next carefully turn the projector slightly so that the point where the axis of the beam meets the east wall moves away from the corner. That part of the image on the north wall will now be a hyperbola (Figure 2.5(d)), while that on the east wall will be part of an ellipse.

2.3   RELEVANT PROPERTIES OF CONIC SECTIONS

Before going on to the projects on orbits it is necessary to describe some of the terms used in connection with conics, and also some of the astronomical terms associated with orbital motion.

An alternative way to trace out a conic, whether it is an ellipse, parabola or hyperbola, is to take a point S, called a focus, and a line, called the directrix, and allow a point P to move in such a way that the ratio $SP:PM$ always has the same value e no matter where P is (Figure 2.6). e is called the eccentricity of the conic.

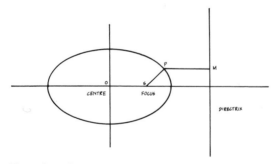

Figure 2.6   Conic sections

(a) The ellipse

If $e < 1$, the conic section traced out by point P will be an ellipse. $A'A$ is called the *major axis of the ellipse*, and $A'O = OA = a$, the *semi-major axis*. Similarly, $OB = b$ is called the *semi-minor axis*. a and b are connected by the relationship

$$b^2 = a^2(1-e^2)$$

If the body P, such as a planet, moves in an ellipse round the Sun as the focus S, it is nearest the Sun when it is at A. This position is called *perihelion*, and

$$AS = a(1-e)$$

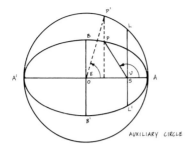

Figure 2.7   Ellipse – true and eccentric anomalies

LL' is called the *latus sectum*

Similarly, the position A' is called *aphelion*, and

$$A'S = a(1+e)$$

Astronomical distances are often measured in astronomical units (a.u.). The semi-major axis of the Earth's orbit round the Sun is taken as the astronomical unit and is 1.5×10^{11} metres.

If the body, such as an artificial satellite or the Moon, moves round the Earth as a focus of the ellipse, the nearest and furthest positions of the satellite from the centre of the Earth are referred to as *perigee* and *apogee* respectively.

In Figure 2.7 if the body is orbiting in an anti-clockwise direction, the angle *ASP* is called the *true anomaly v* for that position of the body.

If a circle is drawn with A'A as diameter, and the line through P perpendicular to A'A meets this circle in P', then the angle AOP' is called the *eccentric anomaly E* of the body. The circle is known as the *auxiliary circle*.

Bodies travelling in elliptical orbits will return to the same position after a time known as *the period*.

(b)   The parabola

If $e = 1$,   the conic section will be a parabola.

Referring to Figure 2.8 it will be seen that the terms major axis and minor axis have no meaning.

The perihelion distance *AS* may be given the symbol *a* or sometimes *q*.

As before, the true anomaly is the angle *ASP*.

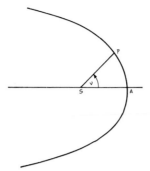

Figure 2.8 Parabola – true anomaly

(c) The hyperbola

If $e > 1$ the conic will be a hyperbola (Figure 2.9).

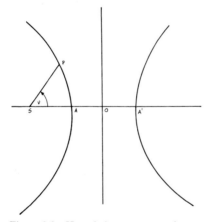

Figure 2.9 Hyperbola – true anomaly

A hyperbola has two branches, but only one portion is of interest here since it is the path of an astronomical body which does not transfer from one branch to the other.

Again, the terms major axis and minor axis have little meaning in this application.

The perihelion distance $AS = a(e - 1)$.

The true anomaly is defined as before as the angle ASP.

Bodies travelling in parabolic or hyperbolic orbits do not return to the same position and the term period is not relevant in these cases.

In all cases of bodies moving in orbits, the distance SP is called the radius or the radius vector.

2.4 PROJECT 1

(i) To construct an ellipse, given the lengths of the major and minor axes.

(ii) To construct a parabola, given the vertex, the direction of the axis and the position of another point known to be on the parabola.

(iii) To construct a hyperbola, given the length of the transverse axis, the direction of the axis and the position of a point known to be on the hyperbola.

(i) Let C be the centre of the ellipse required. Draw a horizontal line $A'CA$ such that $A'C = CA$ = semi-major axis of the ellipse, say 80 mm in this project. Draw a vertical line BCB' such that $BC = CB'$ = semi-minor axis of the ellipse, say 60 mm in this project. Draw circles each with centre C and radius CA and CB respectively.

Draw any radius from C to cut the smaller circle in D and the larger circle in E. Draw a vertical line through E and a horizontal line through D to intersect in P. Then P is a point on the ellipse.

Repeat for other radii, and join the points such as P by a smooth curve to form the ellipse.

(ii) Mark the vertex A and draw the axis of the parabola, AX. From the other given point B, say 100 mm along the axis from A and 60 mm from the axis, draw a line BNB' at right angles to the axis such that $BN = NB'$. Draw lines from B and B' parallel to the axis to cut a line through A at right angles to the axis in D and D' respectively.

Divide BN into say four equal parts, $B1, 12, 23, 3N$. Divide BD into the same number of equal parts $B1, 12, 23, 3D$. Join A to 1 on BD. From 1 on BN draw a line parallel to the axis to cut the last line drawn in P.

Then P is a point on the parabola.

Repeat for the other points 2 and 3, and join the derived points such as P with a smooth curve which will be the parabola required.

For a more accurate result the number of divisions may be increased.

(iii) Draw a horizontal line to represent the transverse axis of the hyperbola. Mark the vertex A and the centre O of the hyperbola to the right of A. In this example we shall take AO to be 50 mm, the distance AO corresponding to the semi-major axis of the hyperbola.

Mark the point B on the paper, to represent the other given point. Again, in this project we shall assume that B is 60 mm to the left of A, and 90 mm above the axis OA. Drop the perpendicular from B on to the axis OA to cut it in N. Complete the rectangle $ANBD$, D being the fourth corner.

Divide BD into a number of equal parts, in our case six, so that we have $B1$, 12, 23, 34, 45, $5D$. Similarly divide BN into the same number of equal parts, so that we have $B1$ 12, 23, 34, 45 and $5N$.

Join O to 1 on BN and join A to 1 on BD. These lines intersect in a point on the hyperbola.

Join O to 2 on BN and join A to 2 on BD giving another point on the hyperbola and so on. Repeat similarly on the other side of OA.

Join up the points of intersection with a smooth curve, which will be the required hyperbola.

2.5   KEPLER'S LAWS OF PLANETARY MOTION

To an observer on Earth the planets appear to move against the background of stars. This motion seems to be very slow and there are peculiarities in the apparent motions which make it very difficult to determine any laws which govern them. However, towards the end of the sixteenth century Tycho Brahe made extensive observations of the planets, particularly the motion of Mars. Study of these observations permitted Johannes Kepler, early in the seventeenth century, to publish three laws of planetary motion. About half a century later Isaac Newton put forward his law of gravitation from which Kepler's three laws can be deduced, the third law in a more exact form than that propounded by Kepler.

2.6   KEPLER'S FIRST LAW

Each planet moves round the Sun in an ellipse, the Sun being at one focus of the ellipse.

This can be generalised into "any astronomical body, moving under the gravitational influence of a second body, traces out a conic with the second body at one focus of the conic".

2.7 KEPLER'S SECOND LAW

When an astronomical body moves under the gravitational influence of a second body we have seen from Kepler's first law that the orbit of the first body is either an ellipse, a parabola or a hyperbola.

The force of attraction is not constant, but varies inversely as the square of the distance between the two bodies. Consequently, since this distance is changing all the time, the velocity of the orbiting body along its path also is not constant, being greatest when nearest the second body (in the case of the Sun at the focus, this will be at perihelion), and least when furthest from the second body.

Kepler found, however, that the rate at which the radius joining the two bodies sweeps out area is constant. Kepler's second law thus states that "the line joining the two bodies sweeps out equal areas in equal times".

Project 2 is designed to verify this as nearly as a graphical method will allow.

2.8 JULIAN DATE

This method of specifying a date is convenient when time intervals have to be determined, as in the case in Project 2. The system was devised by Joseph Scaliger in the sixteenth century. He took as his datum the year 4713 B.C. which, he considered, was earlier than any historical event which could be accurately dated. Thus, all dates specified by this system will be positive, and time intervals are obtained easily by the simple subtraction of two Julian dates. This is much more convenient than using calendar dates. Intervals between two calendar dates are complicated by the differing number of days in the months and also, for longer intervals, by the leap year.

The name Julian was given to the system because the Christian name of Scaliger's father was Julius. It has nothing to do with the Julian calendar.

2.9 PROJECT 2

(i) To construct the orbit of a comet, given the relevant elements of the orbit.

(ii)   To mark the position of the comet at specified dates.
(iii) To verify Kepler's second law of planetary motion, namely that the radius vector sweeps out equal areas in equal times.

Data

The data below gives the relevant elements of the orbit of a comet named P/Wirtanen (P stands for periodic), and also the positions of this comet at specified dates. Hence the orbit must be an ellipse.

Semi-major axis   3.54 a.u.
Eccentricity   0.54

TABLE 2.1

Calendar date	Julian date	Eccentric anomaly
A 1961 July 30	243 7510	31.9°
B 1961 November 27	243 7630	60.4
C 1963 January 11	243 8040	120.7
D 1963 May 11	243 8160	134.1
F 1964 August 23	243 8630	180.7
G 1964 December 21	243 8750	192.3

First calculate the length of the semi-minor axis, using

$$b^2 = a^2(1-e^2)$$

Set out the major and minor axes to a scale of   10 mm = 0.5 a.u.   Draw the circles on the major and minor axes as diameters.

Using the method of Project 1 locate points on the elliptical orbit, and draw the orbit as a smooth curve passing through these points. Using the values of eccentric anomaly given in Table 2.1, plot the position of the comet for each of the dates specified. Note carefully from Figure 2.7 the method needed to do this. Locate the position of the Sun at one focus of the ellipse by measuring a distance ae from the centre of the ellipse (or $a(1-e)$ from one end of the major axis). Draw the radius vector from the Sun to each of the positions of the comet.

The dates have been selected so that the time taken to travel from position A to position B is the same as that to travel from C to D, and the same as from F to G. In each case the time interval is 120 days. Determine the areas *ASB*, *CSD* and *FSG* where S is the Sun, by counting squares on the graph paper, or by using a planimeter or by any other method.

If Kepler's second law is true, then these areas should be equal.

The data given below allows repetition of this project as applied to an elliptical orbit of larger eccentricity to be carried out. The comet is P/Encke.

Data

Semi-major axis a 2.22 a.u.
Eccentricity e 0.85

TABLE 2.2

Calendar date	Julian date	Eccentric anomaly
A 1967 October 17	243 9780	36.0°
B 1968 February 14	243 9900	98.1
C 1968 June 13	244 0020	121.0
D 1968 October 11	244 0140	143.7
E 1969 May 19	244 0360	180.4
F 1969 September 16	244 0480	200.0

For this project a scale of 10 mm = 0.25 a.u. is suggested.

2.10 PROJECT 3

To verify Kepler's second law of planetary motion when the comet is travelling in a parabolic orbit.

Data

The data below gives the relevant elements of Comet Candy, which was at the time observed travelling in a parabolic orbit. Positions of this comet at specified dates are also given. Note that the true anomaly is given.

Distance from vertex to focus q (perihelion distance) 1.06 a.u.
Eccentricity e 1.0

TABLE 2.3

Calendar date	Julian date	True anomaly
A 1961 February 8	243 7338	0°
B 1961 February 18	243 7348	82.7
C 1961 February 28	243 7358	107.6
D 1961 March 10	243 7368	120.0
E 1961 March 20	243 7378	125.8

For this project a scale of 10 mm = 0.5 a.u. is suggested. The interval between the dates is 10 days.

Draw the axis of the parabola. Locate the vertex near the edge of the paper, and measure a distance q along the axis towards the centre of the paper. This will be the focus and the position of the Sun. Measure a distance of $4q$ from the vertex along the axis in the same direction as S, and mark this point N. From N measure off a distance $4q$ perpendicular to the axis on each side. Let these points be F and G. F and G form two corners of a rectangle, and the vertex is the centre of the opposite side of the rectangle. Using the construction given in Project 1, construct the parabola passing through F, G and the vertex A with focus S.

Measure from AS an angle equal to the true anomaly of the comet on date B. Let this cut the parabola at B. Similarly mark the positions of the comet at dates C, D and E using their true anomalies.

Determine the areas ASB, BSC, CSD and DSE.

These areas should be equal, and since the dates are spaced at equal intervals, Kepler's second law should be verified for the parabolic orbit.

2.11 NEWTON'S LAW OF GRAVITATION

We have already mentioned this law in connection with the generalisation of Kepler's first law.

The law says that any two bodies attract each other with a force which varies as the product of the masses, m_1 and m_2, of the bodies, and inversely as the square of the distance d between the bodies. If the bodies are small, this is an accurate statement. If the bodies are large and spherical, the distance between the bodies is the distance between their centres.

Thus
$$F \propto \frac{m_1 m_2}{d^2}$$

or,
$$F = G\frac{m_1 m_2}{d^2}$$

where G is a constant of proportionality known as the constant of gravitation.

If the masses are expressed in terms of the mass of the Sun, and the distance is expressed in astronomical units, the value of G is 0.000 189 approximately. If we work in S.I. units the value of G is 6.673×10^{-11} kg^{-1}m^{3}s^{-2}.

The mass of the Sun is 1.990×10^{30} kg.

2.12   KEPLER'S THIRD LAW OF PLANETARY MOTION

The third of Kepler's laws was published in 1618. It relates the periods of the planets to their semi-major axes. It states that "the squares of the periods of revolution of the planets round the Sun are proportional to the cubes of the corresponding semi-major axes".

If Kepler's third law is true, then

$$a^3 \propto P^2$$

or, $a^3 = K P^2$ where a = semi-major axis
P = period
K = a constant

Taking logarithms of each side,

$$3 \log a = \log K + 2 \log P$$
or, $$\log a = \tfrac{1}{3} \log K + \tfrac{2}{3} \log P$$

This takes the form

$$y = c + n.x$$ where $y = \log a$
$x = \log P$
$c = \tfrac{1}{3} \log K$
$n = \tfrac{2}{3}$

Thus if we plot for the planets $\log a$ as ordinate against $\log P$ as abscissa we should obtain a straight line of slope $\tfrac{2}{3}$. The intercept on the $\log a$ axis should be the value of $\tfrac{1}{3} \log K$.

Sir Isaac Newton's law of gravitation shows that Kepler's third law is not exact, but also depends on the mass of the planet compared with that of the Sun. Even the most massive planet, Jupiter, has a mass of less than one thousandth of that of the Sun, and Kepler's law can be regarded as a sufficiently accurate statement for the present graphical methods.

The constant $K = G.M/4\pi^2$, where G is the gravitational constant, and M is the mass of the Sun.

The true law is $a^3 \propto P^2(M + m)$ where m is the mass of the planet.

2.13 PROJECT 4

(i) To verify Kepler's third law of planetary motion, namely that the squares of the periods of the planets are proportional to the cubes of their semi-major axes.
(ii) To determine the constant of gravitation.

Data

The Table below gives the semi-major axes and the periods of the planets.

TABLE 2.4

Planet	Semi-major axis a (a.u.)	Period P (years)
Mercury	0.39	0.24
Venus	0.72	0.62
Earth	1.00	1.00
Mars	1.52	1.88
Jupiter	5.20	11.87
Saturn	9.58	29.65
Uranus	19.14	83.74
Neptune	30.19	165.95
Pluto	39.44	247.69

The mass of the Sun is 1.990×10^{30} kg.

From the data table, deduce the semi-major axes in metres and the periods in seconds. For this take the astronomical unit as 1.5×10^{11} metres. Draw up a table of $\log_{10} a$ and the corresponding $\log_{10} P$.

Using a scale of 10 mm = 1 unit of $\log_{10} a$ and 10 mm = 0.5 units of $\log_{10} P$, plot $\log_{10} a$ as ordinate against $\log_{10} P$ as abscissa for all the planets.

Draw a straight line through the points and produce it to cut the log a axis at Z.

Read off the intercept between the origin and Z, say c. Equate this to

$$\tfrac{1}{3} \log_{10} \left(\frac{G.M}{4\pi^2} \right)$$

Substituting the value of the mass M of the Sun given in the data, find the value of G, the gravitational constant.

Compare this with 6.67×10^{-11} kg^{-1}m^{3}s^{-2}.

Measure the slope of the line, and compare this with the value $\tfrac{2}{3}$.

2.14 PROJECT 5

(i) To verify Kepler's third law of planetary motion by an alternative method.
(ii) To determine the constant of gravitation.

From the data table in Project 4, deduce the semi-major axes in metres, and the periods in seconds. For this the value of the astronomical unit (a.u.), namely 1.5×10^{11} metres, will be needed.

Draw up a table of a^3 and the corresponding P^2 for the planets Uranus, Neptune and Pluto only. Because of the wide spread of these values over the whole range of planets it is not possible to include all the planets on one graph. It is, however, possible to repeat the project with three other planets.

Plot a^3 as ordinate to a scale of 10 mm = 10 units of a^3, against P^2 as abscissa to a scale of 10 mm = 5 units of P^2.

Draw the straight line through the points, and note that it passes through the origin. Thus $a^3 \alpha P^2$. Measure the slope of the line. Equate this to $G.M/4\pi^2$, and assuming the mass of the Sun, $M = 1.990 \times 10^{30}$ kg, calculate the value of G, the constant of gravitation. Compare this with the value 6.67×10^{-11} kg^{-1}m^{3}s^{-2}.

2.15 KEPLER'S EQUATION

The position of a planet or a comet in a given elliptical orbit may be obtained for any time t from Kepler's equation, provided that the time of perihelion passage T and the period P of the orbit are known.

From the period, a quantity $(\bar{n})$, known as the mean daily motion, is determined. This is defined as

$$\bar{n} = \frac{360°}{P}$$

the units usually being in degrees per day, so that P must be given in days.

Kepler's equation then states that

$$\bar{n}(t-T) = E - e° \sin E$$

where E is the eccentric anomaly in degrees at time t.

If E can be determined, we have seen in Project 2 how we can locate the position of the orbiting body.

The normal method of solving this equation for E is to obtain an approximate value for E, and by a successive approximation method arrive at a sufficiently accurate value. The approximate value of E may be obtained from a graphical method, and this will be sufficient for our purpose to locate the approximate position of the orbiting body.

We first put $\bar{n}(t - T) = M$, which is called the mean anomaly. Then Kepler's equation can be re-arranged as

$$E - M = e° \sin E$$

If two graphs are plotted,

$$y = E - M$$

and

$$y = e° \sin E$$

then the point where they intersect will give that value of E such that $E - M = e° \sin E$, or that value of E which satisfies Kepler's equation for the time t specified.

2.16 PROJECT 6

To determine the eccentric anomaly E at a specified time for a body travelling in an elliptical orbit, using Kepler's equation

$$\bar{n}(t-T) = E - e° \sin E$$

Data

The comet chosen for this project is that selected in Project 2 so that a check of the result may be made.

Comet P/Wirtanen
Eccentricity of orbit e 0.54
Period P 6.67 years
Time of perihelion passage T 1961 April 15.3
           corresponding to Julian date   243 7404.8

Convert e into degrees by multiplying it by 57.3. Calculate $\bar{n}$ in degrees per day by dividing 360 by P in days. Taking the date 1963 January 11, corresponding to Julian date 243 8040, determine $M = \bar{n}(t-T)$.

Using a scale of   10 mm = 2.5 units   for the y scale as ordinate, and a scale of   10 mm = 10°   for the E scale as abscissa, plot the graph of $y = e° \sin E$.   Intervals of 15° for E will be found suitable.

On the same graph axes draw the graph of $y = E - M$.

Determine the value of E where the two graphs intersect, and compare this with the value given in Project 2 of 120.7°.

2.17   NOMOGRAPHS

Where the same problem must be solved repeatedly using different data it is often convenient to construct a nomograph or alignment chart. Use of such a nomograph produces a quick, albeit approximate, solution. One such problem is to determine the semi-major axes of say, artificial satellites once their periods of revolution round the Earth have been measured. Another is the conversion of the position of an astronomical body from one set of co-ordinates to another.

A number of nomographs useful in astronomy are given throughout this book. For readers who require a general mathematical theory on which nomographs are based, this is provided in Appendix A1.

2.18   AN ASTRONOMICAL NOMOGRAPH

A nomograph relating the periods of planets to the semi-major axes of their orbits will now be devised using simple reasoning.

Since from Kepler's third law

$$a^3 \propto P^2$$

then $$a^3 = k^2.P^2$$

where k^2 is a constant which we have already met. It is equal to $G.M/4\pi^2$, where G is the universal gravitational constant, and M is the mass of the Sun. The constant is written as k^2 rather than k since we shall require to write the square root of the constant rather than the constant itself.

Writing the last equation

$$a^3 = (kP)^2, \quad \text{and taking logarithms,}$$

we get $$3 \log a = 2 \log (kP) \quad \text{where } k = \sqrt{G.M/4\pi^2}$$

or $$\frac{\log a}{\log (kP)} = \frac{2}{3}$$

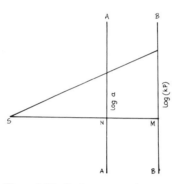

Figure 2.10 Basic nomograph

In Figure 2.10 let AA be the line on which values of the semi-major axis a can be placed on a logarithmic scale. Let BB be the line on which values of the product kP can be placed, also on a logarithmic scale. Let S represent the Sun.

Then, by similar triangles

$$\frac{\log a}{\log kP} = \frac{SN}{SM}$$

If the lines are spaced from S such that the ratio $SN/SM = \frac{2}{3}$,

then $$\frac{\log a}{\log kP} = \frac{2}{3}$$

and the quantities on the two lines are related by Kepler's third law. If E and E' represent corresponding values of a and P for the Earth, and the scales AA and BB are moved vertically relative to one another until SEE' is a straight line (Figure 2.11), then it follows that a straight

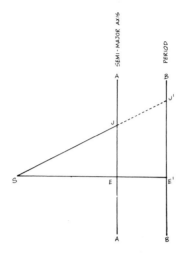

Figure 2.11   Adjusted nomograph

line from S to J, the value on line AA for Jupiter, must, when produced, pass through J', the correct value for Jupiter on line BB.

It is often possible to combine several nomographs into a single nomograph linking several quantities involved.

2.19   PROJECT 7

To construct a nomograph relating the semi-major axes of the orbits of the planets and their orbital periods.

Data

The Table below gives the semi-major axes of the planets in astronomical units, and also the period, in years, for the orbits of the Earth and Jupiter.

If we take the unit of distance as the astronomical unit, the unit of mass as that of the Sun and the unit of time as the day, then the value of k is very close to 0.0172.

TABLE 2.5

Planet	Semi-major axis (a) a.u.	Period (P) years
Mercury	0.39	
Venus	0.72	
Earth	1.00	1.00
Mars	1.52	
Jupiter	5.20	11.87
Saturn	9.58	
Uranus	19.14	
Neptune	30.19	
Pluto	39.44	

Taking the values in the given data table, calculate $\log_{10} a$ for each of the planets listed. Taking the values of P in the data table, convert these to days given that 1 year is 365.25 days. Calculate $\log_{10} kP$ for the Earth and for Jupiter.

On a piece of graph paper, mark a point S near to the left hand edge and 180 mm from the bottom of the paper. 40 mm from S measured horizontally draw a vertical line on which to represent values of the semi-major axes. 60 mm from S measured horizontally, draw a second vertical line on which to represent values of the periods. Draw SEE' horizontally to cut the two lines in E and E' respectively.

Since for the Earth, $a = 1$ a.u. and therefore $\log_{10} a = 0$, the position of E may be taken as the datum from which the other values of $\log_{10} a$ may be measured. Mark the position J of Jupiter on the first vertical line, to a scale of 20 mm = 1 unit of $\log_{10} a$.

Since $\log_{10} kP$ for the Earth is 0.798, the datum O for the second vertical line will be 16.0 mm below E', using the same scale as for the first line. From O measure a distance OJ' on the second vertical line equivalent to $\log_{10} kP$ for Jupiter.

S, J and J' should then be in a straight line.

The two scales can now be marked in a.u. and days respectively. On the first line the point S will be marked 1 a.u. A point 20 mm above E will represent 1 logarithmic unit and will therefore be marked 10 a.u. Similarly, a point 20 mm below E will be marked 0.1 a.u. On the second line distances are proportional to $\log_{10} kP$. Since $\log_{10} kP = \log_{10} k + \log_{10} P$, the effect of k can be removed from the scale by measuring a distance $\log_{10} k = -1.76$ (-35.2 mm) from O to a point C below O, and then taking C as reference for a scale of $\log_{10} P$. Then E' which is now ($16.0 + 35.2$ mm) above C will be marked 365.25 days since $\log_{10} 365.25 = 2.56$ (51.2 mm). This is more conveniently marked 1 year. The point on the second vertical line 20 mm above E' will be marked 10 years and the point 20 mm below E' will be marked 0.1 years. Finer graduations on each scale can be marked on the basis that $\log_{10} 2 = 0.301$ (6.0 mm above E or E'), $\log_{10} 3 = 0.477$ (9.5 mm above E or E') and so on. The nomograph is now calibrated and ready for use.

Using the values of a from the data table, mark the positions of the planets on the first vertical line. Draw a straight line from S through each planet position and produce it to cut the second vertical line. Read off the period of each planet and compare the values obtained with those given in Project 4.

2.20 PROJECT 8

To extend the nomograph in Project 7 so that it will relate the periods and the semi-major axes of satellites, either natural or artificial, revolving round planets or the Moon.

Data

TABLE 2.6

Parent body	Mass of parent body relative to the Sun	Satellite	Semi-major axis (km)	Period
Jupiter	9.55×10^{-4}	Io	421 600	1.77 days
Earth	3.01×10^{-6}	Artificial	7 378 (circular orbit)	1 h 34 min
Mars	3.22×10^{-7}	Deimos	23 500	1.26 days
Saturn		Rhea	526 700	4.52 days
Uranus		Oberon	585 960	13.46 days
Moon	3.70×10^{-8}	Command and service module (C.S.M.)	1 838	

Using the nomograph of Project 7, draw a vertical line through S. This line is to represent the masses of the parent bodies relative to the Sun. Taking the mass of the Sun at S as 1.0, mark in masses of 10^{-1}, 10^{-2}, 10^{-3}, 10^{-4}, 10^{-5}, 10^{-6}, 10^{-7} and 10^{-8} of the Sun's mass to a logarithmic scale. If we use the same scale as before, namely 20 mm = 1 logarithmic unit, these will be respectively 20 mm, 40 mm, 60 mm . . . 160 mm below S. On the semi-major axis scale mark in distances of 10^7, 10^6, 10^5, 10^4, 10^3 and 10^2 km taking the astronomical unit as 1.5×10^8 km.

On the period scale mark in times of 100, 10 and 1 day and 10, 5, 4, 3, 2, 1.5 and 1 hour, 50 and 40 minutes. Remember that the zero for mass is at S, the zero for distance is E but the datum for period is at C.

Draw a straight line from $P = 1.77$ days to $a = 421600$ km and produce this to cut the vertical through S in J. This should read 9.55×10^{-4} of the Sun's mass. Repeat for the values of P and a given for Earth and Mars, cutting the mass axis in 3.01×10^{-6} and 3.22×10^{-7} of the Sun's mass respectively.

We now have enough confidence that the construction of the mass scale was correct and can therefore use it similarly to determine the position (and mass) of Saturn and Uranus respectively.

In reverse we can obtain the period of revolution of a C.S.M. flying 100 km above the surface of the Moon in a circular orbit. The Moon has a radius of 1738 km.

To establish the correctness of the nomograph readers should refer to Appendix A2.

2.21   VELOCITY OF ESCAPE

If we project an object vertically upwards at a relatively low velocity, the attraction of the Earth will cause the object to return to the surface of the Earth. If we increase the velocity of projection, the object will rise to a greater height before beginning its return. Continued to its limit, there will be a velocity which will cause the body to travel an extremely large distance away from the Earth, at which point the gravitational pull of the Earth on the body has diminished to zero. There will thus be no tendency for the body to return to the Earth, and we say that the body has escaped.

We refer to the velocity necessary to achieve this as the velocity of escape. The extremely distant position achieved we refer to as infinity.

With the advent of space travel, it is necessary to give a rocket a velocity larger than the velocity of escape if the satellite is not to orbit the Earth. From the astronomer's point of view the concept of velocity of escape is important in explaining why planets, or natural satellites of planets, lack certain gases in their atmospheres, or even lack atmospheres at all.

Gas molecules move about with a velocity which depends on the temperature of the gas. If the temperature of the atmosphere of a planet becomes high enough, the velocities of the gas molecules in the original atmosphere may exceed the escape velocity for the planet. After a time then, these gases will escape into space. For instance, on Mars where the Viking I spacecraft shows that the temperature reaches 240 K, one would expect no molecular hydrogen to be found in the lower atmosphere since the mean velocity of hydrogen molecules at this temperature is very close to the escape velocity of about 5 km/s. In fact the amount of hydrogen present is below the threshold of detection of the Viking instruments. It is thought that the chemical processes in the Martian atmosphere produce both molecular and atomic hydrogen and indeed, Mariner spacecraft during their orbits of Mars have detected small amounts of atomic hydrogen in the upper Martian atmosphere.

2.22   GRAPH OF ACCELERATION OF A BODY PLOTTED AGAINST DISTANCE

If a body is moving with constant acceleration f over a distance s, its motion can be represented graphically as in Figure 2.12. The area of this graph will be fs.

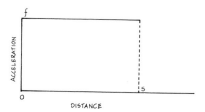

Figure 2.12   Distance-acceleration graph

For such a motion in which the initial velocity is u and the final velocity v, many readers will be familiar with the relationship between these quantities,

$$v^2 = u^2 + 2f.s$$

If, as a special condition, the body starts from rest, that is $u = 0$, the final velocity will be given by

$$v = 2f.s$$

In reverse, if the body starts with a velocity v and decelerates to zero, the relationship $v^2 = 2f.s$ still holds, and the square of the velocity of projection is twice the area under the graph.

The same is equally true if the acceleration or deceleration is not constant, as in the case of a body projected from the Earth. Here the acceleration due to gravity is decreasing as the distance from the Earth increases.

Mathematically minded readers will recognise that

$$f = v.\frac{dv}{ds}$$

Integrating both sides with respect to s between the limits of travel s_1 and s_2,

$$\int_{s_1}^{s_2} f.ds = \int_u^v v.\frac{dv}{ds}.ds = \int_u^v v.dv$$

Hence,

$$\text{area under curve} = \left[\frac{v^2}{2}\right]_u^v = \frac{v^2}{2} - \frac{u^2}{2}$$

Since the body has to end up with zero velocity $(u = 0)$,

$$\text{area under curve} = \frac{v^2}{2}$$

or,
$$v^2 = 2(\text{area under curve of } f{-}s)$$

2.23 VARIATION OF ACCELERATION DUE TO GRAVITY

If a body is released near to the surface of the Earth, it will fall freely towards the Earth with an acceleration of about 9.81 m/s^2. Due to the non-uniformity of the Earth, this value varies slightly from place to place.

The acceleration is explained by Newton's first law of motion, which, in effect, states that if a force acts on a body, it must accelerate. The force involved here is the force of gravitational attraction between the Earth and the body. Newton's second law does, in fact, imply that the force acting is equal to the product of the mass of the accelerating body and its acceleration.

Thus, if the body is at radius r from the centre of the Earth, the force acting on it is, by the law of gravitation

$$F = G.\frac{Mm}{r^2} \quad \text{where} \quad \begin{aligned} M &= \text{mass of the Earth} \\ m &= \text{mass of the body} \end{aligned}$$

But, by Newton's second law of motion,

$$F = m.g_r \quad \text{where} \quad g_r = \text{acceleration due to gravity at radius } r$$

Hence

$$m.g_r = \frac{G.Mm}{r^2}$$

or,

$$g_r = \frac{G.M}{r^2}$$

$$= \frac{\text{constant}}{r^2}$$

In particular, if g is the acceleration due to gravity at the surface of a planet of radius R, then the acceleration due to gravity at a distance r from the centre of the planet is given by

$$g_r = \frac{R^2}{r^2}.g$$

In this we assume that r is greater than R, that is the body being considered is outside the planet itself.

2.24 PROJECT 9

(i) To show how the acceleration due to gravity varies with distance from a planet.
(ii) To determine the velocity of escape from the planet.

Data

The figures given below refer to conditions at the equator of each planet:

Acceleration due to gravity at the surface of the Earth	$= 9.78 \text{ m/s}^2$
Acceleration due to gravity at the surface of Mars	$= 3.84 \text{ m/s}^2$
Radius of the Earth	$= 6380 \text{ km}$
Radius of Mars	$= 3330 \text{ km}$

Using the formula $g_r = R^2/r^2 \times g = 6380^2/r^2 \times 9.87$ m/s^2, calculate the acceleration due to gravity at distances from the centre of the Earth of 6380, 7000, 8000, 10 000, 15 000, 20 000, 25 000, 30 000, 40 000, 50 000, 60 000 and 100 000 km.

Using scales 10 mm = 5000 km for distance and 10 mm = 1 m/s^2 for acceleration, plot a graph of gravitational acceleration as ordinate against distance from the centre of the Earth as abscissa. By counting squares or otherwise, measure the area between the distance 6380 km and 100 000 km. Strictly speaking, the area should be measured to a much greater distance than this, but the values of g_r are getting impractically small.

Multiply the area obtained by two, and convert to velocity squared units as follows:

A square 10 mm $\times$ 10 mm represents $\dfrac{1}{1000} \times 5000 = 5$ km^2/s^2.

Thus this factor times twice the measured area represents the square of the velocity of escape. Calculate this value and take the square root. Compare the derived value with the value of 11.1 km/s, which is the correct value of the velocity of escape from the Earth.

This project may be repeated for Mars (or indeed any other planet or the Moon).

For Mars, use scales 10 mm = 2000 km for distance and 10 mm = 0.25 m/s^2 for acceleration. Compare the value obtained for the velocity of escape from Mars with the correct value of 5.1 km/s.

3

The Moon

3.1 THE MOON'S ORBIT

The Moon travels in an orbit round the Earth, its parent body. This orbit is complex because of the additional influence of the Sun and of the other planets. As a first approximation, however, it may be said that the orbit is an ellipse with the centre of the Earth at one focus. The eccentricity e of the ellipse is small as may be seen by inspection of the column in Table 3.1 headed Diameter. This gives the angle θ', which can be measured, subtended at the observer by the true diameter of the Moon for dates throughout one orbit. The variation is small, which means that the orbit is nearly circular and therefore that the eccentricity is small.

If the plane of the Earth's orbit round the Sun is taken as reference, the plane of the orbit of the Moon is tilted at about 5° to the reference plane. For the purposes of the next project it will be assumed that the reference plane and the plane of the Moon's orbit are the same. This will account for some small error in Project 10.

The Longitude column in Table 3.1 gives the angle θ which the radius from the Earth to the Moon makes with a line drawn in a fixed direction, namely towards γ the First Point of Aries (see Section 4.1). The angle a between the direction of perigee and the direction of γ will be needed in Project 10.

As the Moon progresses round its orbit, its distance from the Earth will change and, as we have said, the angle θ' subtended at the observer by the diameter of the Moon will change. Since the distance r of the Moon from the Earth is large compared with the Moon's diameter we can apply $\text{arc} = r.\theta'$, where, in this case, the arc can be taken as the linear diameter D of the Moon and θ' is in radians.

Thus $r = D/\theta'$

We do not suppose in Project 10 that we know the true linear diameter of the Moon, but we take note of the fact that it is of constant value. Also the angles measured will be in minutes of arc but the conversion to radians is also achieved by multiplying by a constant factor.

Hence we may write $r' = \text{constant}/\theta'$

Thus values of r' calculated for measured θ' would enable us to plot the true shape of the Moon's orbit to a scale which will depend on the chosen value of the constant. We shall not, however, do this but we shall use the above ideas to determine the eccentricity of the Moon's orbit. A suitable value for the constant for our purposes is 100.

3.2 PROJECT 10

To determine the eccentricity of the Moon's orbit

Data

The Table below gives the diameter θ' of the Moon as seen from the Earth on specified dates throughout a lunar month. The Table also gives corresponding longitudes θ referred to the plane of the Earth's orbit as described above.

The polar equation of the elliptical orbit, taking the Earth as origin, is

$$r = l/[1 + e.\cos(\theta - a)]\ \text{where}\ l = \text{semi-latus sectum}$$
$$\text{(See Section 2.3)}$$

This can be written

$$r = l[1 + e.\cos(\theta - a)]^{-1}$$

Since e is known to be small and $\cos(\theta - a)$ has a maximum of only unity we can apply the binomial theorem, giving the close approximation

$$r = l[1 - e.\cos(\theta - a)]$$

Applying the constant as described in Section 3.1

$$\frac{100}{\theta'} = \frac{100}{\theta'_l} - \frac{100}{\theta'_l}.e.\cos\varphi$$

where θ_l', is the angular diameter of the Moon when it is at distance l from the Earth, and φ is $(\theta - a)$ for each position of the Moon in the Table.

Thus if we plot $100/\theta'$ as ordinate to a scale of 20 mm = 0.1 units against $\cos \varphi$ as abscissa to a scale of 10 mm = 0.1 we should obtain a straight line of slope $-100.e/\theta_l'$. The line will cut the vertical through $\cos \varphi = 0$ ($\varphi = 90°$ and $270°$) at a value which is $100/\theta_l'$.

TABLE 3.1

Date	Diameter	Longitude
1961 July 4.0	32.20'	356° 58'
July 6.0	31.44	24 25
July 8.0	30.74	50 42
July 10.0	30.16	76 00
July 12.0	29.74	100 32
July 14.0	29.48	124 29
July 16.0	29.44	148 05
July 18.0	29.50	171 42
July 20.0	30.16	195 53
July 22.0	31.00	221 13
July 24.0	32.00	248 14
July 26.0	32.94	277 04
July 28.0	33.40	307 09
July 30.0	33.20	337 22
July 31.0	32.86	352 10
Aug. 1.0	32.44	6 37

From the Table calculate $100/\theta'$ where θ' is the given angular diameter of the Moon in minutes of arc.

Subtract the longitude at perigee a (corresponding to the greatest value of θ') from each of the longitudes given to obtain the corresponding values of φ. Then plot $100/\theta'$ against $\cos \varphi$ to the scales given above.

Measure the slope of the graph and read the intercept $\dfrac{100}{\varphi_l'}$ on the $\cos \varphi = 0$ line.

Then from

$$m \text{ (the slope)} = -\frac{100}{\theta_l'} . e$$

obtain

$$e = -\frac{m.\theta_l'}{100}$$

Compare this value with 0.055 which is the eccentricity of the Moon's orbit.

3.3   THE PHASES OF THE MOON

Even the most casual of observers will have noticed that the shape of the Moon, as seen from the Earth, changes considerably with time. Sometimes the Moon appears as a crescent, sometimes as a disc and at other times as a shape between these two extremes.

The phases of the Moon, as they are called, arise from the changing relative positions of the Earth and the Moon. The Sun always illuminates half of the sphere of the Moon, but the Moon will only appear full if all the illuminated portion can be seen from the Earth. This means that the Earth lies between the Moon and the Sun.

If the Earth lies on the other side of the Moon from the Sun, no portion of the illuminated Moon can be seen, and the Moon is new.

In other positions of the Moon relative to the Earth, only parts of the Moon will appear illuminated (Figure 3.1).

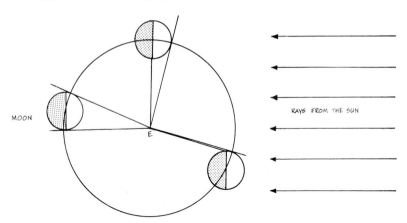

Figure 3.1   Phases of the Moon

Project 11 shows how to construct the Moon as it would appear from the Earth for any chosen relative position of the Earth and Moon.

3.4   PROJECT 11

To construct the shape of the Moon as seen from the Earth for various positions of the Moon relative to the Earth.

Take a position of the Moon relative to the Earth as shown in Figure 3.2 For the accuracy obtainable in drawing, the rays of light from the Moon to the Earth may be regarded as parallel, in view of the large distance of the Moon from the Earth.

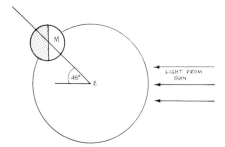

Figure 3.2   Moon's phase

Draw two circles side by side, each 50 mm diameter, with centres A and B respectively 95 mm apart. Draw a line AC such that $\angle BAC$ is 135°. Locate C 65 mm from A. With centre C draw a third circle of the same size as the others. Draw a line DCD' through C perpendicular to AC. Draw a line FAF' through A perpendicular to AB and a line GBG' through B perpendicular to AB. With centre A, draw concentric circles reasonably equally spaced within this circle. About six circles will be enough.

If we regard the circle with centre A as a plan view of the Moon, that is a view at right angles to the plane of the Moon's orbit, then these circles represent different levels on the surface of the Moon above that plane. Let one of these circles cut AF in X. From X draw a line XX' parallel to AC to cut DCD' in M. From X draw another line parallel to AB to cut the circle with centre B in Z and Z' and the line GBG' in Y.

Then YZ represents the height of X above the plane of the Moon's orbit. Mark off distances MR and MR' equal to YZ on each side of DCD' along the line XX'.

Repeat for the other circles, and join up the points similar to R and R' with a smooth curve.

This curve will be the shape and position of that edge of the Moon which appears to move across the Moon throughout the lunar month, that is the terminator. The other edge will be the semi-circular edge through D', that is the limb.

This project may be repeated for other angles for $\angle BAC$.

Figure 3.3   Moon craters, mountains and maria   *Photo © Lick Observatory, University of California,*
Santa Cruz, U.S.A.
Reproduced by kind permission

3.5   THE HEIGHT OF LUNAR CRATER WALLS

The photograph (Figure 3.3) shows that the surface of the Moon is covered with approximately circular craters. There are also other rugged features which are the mountains and mountain ranges of the Moon. Apart from the interest which astronomers have in the Moon as a whole, estimation of the size of the mountains and craters is of concern to astronauts. Until the heights of crater walls could be measured from the surface of the Moon using direct methods, indirect methods had to be applied. Project 12 illustrates one of them.

On Earth we can quote the heights of mountains from a common datum, namely the level of the sea. Since there are no seas on the Moon (the so-called marias are not seas, but relatively flat areas of land) it is only possible to quote the height of mountains or crater walls above the surrounding neighbouring surface. Some craters appear to have been partially filled in, and the height of the walls, as measured from outside may be larger, or smaller, than that from inside the walls.

There are a number of ways of determining the approximate height of crater walls. The one described here depends on measuring the length of shadow cast by the wall of the crater.

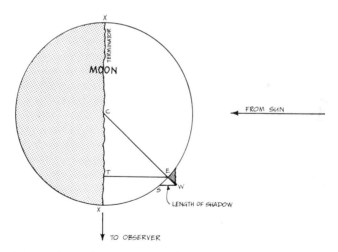

Figure 3.4   Height of lunar mountain

Referring to Figure 3.4, *EW* is the crater and *WS* is the shadow which it casts, the Sun being on the right of the Moon. The illuminated portion

Figure 3.5   Moon at first quarter

of the Moon will be that half to the right of XX, the terminator. Figure 3.4 and Figure 3.5 refer to the situation when the Moon is 7 days old, or the phase is the first quarter.

If $l = WS$ = the length of shadow (mm)
  $d = ET$ = the distance of the crater wall from the terminator (mm)
  $R = \frac{1}{2}XX$ = the radius of the Moon (mm)
  $h = EW$ = the height of crater wall (mm)

all measured on the photograph, then triangle CTE is similar to triangle SEW, and

$$\frac{SW}{CE} = \frac{WE}{ET}$$

or
$$\frac{l}{R} = \frac{h}{d}$$

or height of crater wall, $h = \dfrac{l.d}{R}$

This will give the height of the wall in millimeters, and this has to be translated into true distance in metres.

R mm on the photograph corresponds to 1738 km, the actual radius of the Moon. Hence

$$\text{actual height of wall} = \frac{l.d}{R} \times \frac{1738}{R} \times 1000 \text{ metres.}$$

The crater wall has been shown on the edge of the Moon in Figure 3.4 for clarity of explanation. More accurate results may be expected when the crater is near the terminator when the phase of the moon is first or third quarter, since the shadow length as we see it will be less affected by the curvature of the surface.

3.6   PROJECT 12

To determine the height of the wall of the Moon crater Aliacencis above the level of the floor of the crater.

The photograph (Figure 3.5) shows the Moon at first quarter. From the key map of this half Moon, (Figure 3.6) locate the crater Aliacencis on the photograph.

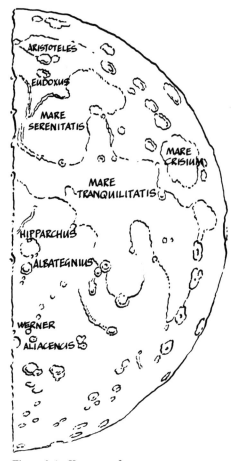

Figure 3.6 Key map of craters

On the photograph measure the dimension YY', the length of shadow of the crater wall (Figure 3.7). Estimate this dimension to one tenth of a millimetre, using a lens if necessary. Measure the distance of Y' from the terminator, also in millimetres. Measure the length of the terminator itself in millimetres, which, since the Moon's phase is first quarter, will also be the diameter of the Moon on the photograph.

Apply the formula derived in the foregoing pages,

$$\text{height of wall} = \frac{l.d}{R} \times \frac{1738}{R} \times 1000 \text{ metres.}$$

Compare your value with 5030 metres.

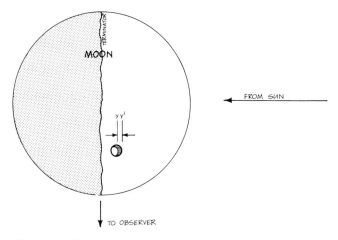

Figure 3.7 Moon crater shadow

Repeat this project with the crater Albategnius, or any other crater near the terminator with a measurable amount of shadow. The crater Albategnius is quoted as having a wall 3850 metres above the floor of the crater. Werner has a wall 4570 metres high.

3.7 HARVEST MOON AND HUNTER'S MOON

Observation of the time at which the Moon comes over the horizon shows that moonrise is delayed by an average of about 50 minutes on each successive night.

However, as we shall see in Project 13, there are a few days in each lunar month during which the delay in moonrise is considerably less than for the rest of the lunar month. This is due to the movement in declination (see Section 4.2(b)), which is effectively its latitude in the sky referred to the plane of the Earth's equator. The delay is least when the rate of increase of the Moon's declination is greatest.

In the month of September, the period of least delay in moonrise coincides with the full Moon, and the time of moonrise coincides with the time of sunset. Thus at this time the loss of light due to the setting Sun is compensated in part by the light from the full or nearly full Moon with the least possible delay. Since this time of the year is harvest time, the Moon is referred to as the Harvest Moon. It clearly allows a longer working day for harvesting.

A similar, if less pronounced, phenomenon occurs at the next full Moon, which is referred to as the Hunter's Moon.

3.8 PROJECT 13

(i) To plot the time of moonrise and the time of sunset throughout the year.

(ii) To observe the time of year when the interval between sunset and moonrise is least on successive nights.

(iii) To determine the approximate dates of the full Moon.

Data

The times of moonrise and of sunset for various dates throughout a year are given in the Tables below.

Because the time interval of approximately one year is so spread out, this project is best carried out on four separate sheets of paper, each representing about three months.

Set up axes, the ordinate representing time of moonrise or time of sunset on the same scale of 24 hours. Use a scale of 10 mm = 2 hours.   The abscissa will represent time of year to a scale of 10 mm = 4 days for a range of about three months.

Using the data in the Tables, plot the times of moonrise against the corresponding dates for three successive lunar months. Join up each set of points with a smooth curve. Similarly, plot the time of sunset against date for the same three months, and join these points with a smooth curve.

Repeat this, on separate sheets, for the remaining data in groups of three months.

Note that in each lunar curve the slope of one portion of the curve is distinctly less than the slope of the remainder of the curve for that lunar month. If the steep portion of the curve cuts the sunset curve, the delay of moonrise on successive nights about that date is relatively large compared with the delay when the less steep portion of the curve cuts the sunset curve. Thus, for the latter case, moonrise follows quickly on sunset for several days running.

The Harvest Moon in September is one such case, and the Hunter's Moon in October another.

TABLE 3.2

Date		Moonrise		Date		Moonrise		Date		Moonrise	
Jan	2	1h	47m	Feb	1	3h	38m	Mar	1	2h	43m
	4	4	37		3	6	10		3	4	52
	6	7	25		5	7	33		5	5	57
	8	9	08		7	8	12		7	6	30
	10	9	54		9	8	37		9	6	55
	12	10	20		11	9	03		11	7	23
	14	10	43		13	9	40		13	8	09
	16	11	14		15	10	51		15	9	36
	18	12	10		17	12	50		17	11	48
	20	13	53		19	15	16		19	14	16
	22	16	14		21	17	42		21	16	41
	24	18	41		23	20	07		23	19	08
	26	21	05		25	22	37		25	21	45
	28	23	31								

TABLE 3.3

Date		Moonrise		Date		Moonrise		Date		Moonrise	
Mar	28	0h	31m	Apr	27	1h	30m	May	25	0h	04m
	30	2	47		29	2	25		27	0	49
Apr	1	4	00	May	1	2	55		29	1	15
	3	4	36		3	3	17		31	1	39
	5	5	01		5	3	48	Jun	2	2	12
	7	5	27		7	4	38		4	3	12
	9	6	08		9	6	11		6	5	02
	11	7	26		11	8	29		8	7	27
	13	9	33		13	10	56		10	9	53
	15	12	00		15	13	20		12	12	15
	17	14	25		17	15	47		14	14	42
	19	16	51		19	18	27		16	17	25
	21	19	27		21	21	17		18	20	11
	23	22	16		23	23	26		20	22	02
									22	22	54
									24	23	22
									26	23	46

TABLE 3.4

Date		Moonrise		Date		Moonrise		Date		Moonrise	
Jun	29	0h	17m	Jul	31	1h	48m	Aug	28	0h	49m
Jul	1	1	11	Aug	2	4	14		30	3	16
	3	2	51		4	6	39	Sep	1	5	40
	5	5	12		6	9	00		3	8	00
	7	7	39		8	11	24		5	10	28
	9	10	01		10	14	01		7	13	05
	11	12	24		12	16	38		9	15	28
	13	14	59		14	18	25		11	16	54
	15	17	45		16	19	17		13	17	37
	17	19	53		18	19	47		15	18	05
	19	20	56		20	20	13		17	18	33
	21	21	29		22	20	47		19	19	14
	23	21	53		24	21	49		21	20	31
	25	22	23		26	23	39		23	22	37
	27	23	11								

TABLE 3.5

Date		Moonrise		Date		Moonrise		Date		Moonrise	
Sep	26	1h	04m	Oct	24	0h	03m	Nov	23	1h	20m
	28	3	28		26	2	26		25	3	43
	30	5	50		28	4	48		27	6	19
Oct	2	8	16		30	7	19		29	8	57
	4	10	53	Nov	1	9	58	Dec	1	10	52
	6	13	19		3	12	10		3	11	51
	8	14	53		5	13	24		5	12	23
	10	15	41		7	14	02		7	12	48
	12	16	10		9	14	29		9	13	18
	14	16	37		11	14	56		11	14	10
	16	17	14		13	15	40		13	15	49
	18	18	23		15	17	03		15	18	13
	20	20	23		17	19	17		17	20	43
	22	22	50		19	21	47		19	23	04

TABLE 3.6

Date		Sunset		Date		Sunset		Date		Sunset	
Jan	0	15h	58m	Feb	4	16h	53m	Mar	1	17h	39m
	5	16	03		9	17	02		6	17	48
	10	16	10		14	17	11		11	17	57
	15	16	18		19	17	21		16	18	06
	20	16	26		24	17	30		21	18	14
	25	16	34						26	18	23
	30	16	43						31	18	31

TABLE 3.7

Date		Sunset		Date		Sunset		Date		Sunset	
Mar	31	18h	31m	May	5	19h	31m	Jun	4	20h	13m
Apr	5	18	40		10	19	39		9	20	18
	10	18	48		15	19	47		14	20	21
	15	18	57		20	19	54		19	20	23
	20	19	05		25	20	01		24	20	24
	25	19	20		30	20	07		29	20	24
	30	19	22								

TABLE 3.8

Date		Sunset		Date		Sunset		Date		Sunset	
Jul	4	20h	22m	Aug	3	19h	47m	Sep	2	18h	46m
	14	20	15		8	19	38		7	18	35
	24	20	03		13	19	29		12	18	23
	29	19	56		18	19	19		17	18	11
					23	19	08		22	18	00
					28	18	57		27	17	48

TABLE 3.9

Date		Sunset		Date		Sunset		Date		Sunset	
Oct	2	17h	36m	Nov	1	16h	33m	Dec	1	15h	53m
	7	17	25		6	16	24		6	15	50
	12	17	14		11	16	16		11	15	49
	17	17	03		16	16	08		16	15	49
	22	16	52		21	16	02		21	15	50
	27	16	42		26	15	57		26	15	53
									31	15	58

At the Harvest Moon, the Moon is at the First Point of Aries (see Section 4.1) and the Sun is diametrically opposite at the Autumnal Equinox. The Moon is therefore full, giving the maximum light assistance to harvesters. Hence the date at which the lunar curve in September crosses the sunset curve will be the date of the full Moon. Read this date and compare with September 15 days 11 hours.

4

(i) The Celestial Sphere
(ii) The Planets

4.1 THE CELESTIAL SPHERE

To an observer on Earth, the sky seems to be part of a sphere, known as the celestial sphere. The distance assumed for the radius of this sphere is not important, but the directions of objects projected on to it from the observer *are* important.

The Earth rotates on an axis through the north and south poles. The plane perpendicular to this axis and passing through the origin is known as the equatorial plane. This plane will cut the celestial sphere in a (great) circle known as the equator.

In the course of a year, the Sun appears to move along a path on the celestial sphere which is not parallel to the equatorial plane, but along a plane known as the ecliptic plane. The ecliptic plane cuts the celestial sphere in another (great) circle known as the ecliptic.

The angle between the equatorial plane and the ecliptic plane is referred

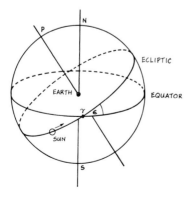

Figure 4.1   Celestial equator and ecliptic

to as the angle of obliquity, which has a value of about $23\frac{1}{2}°$. The equator and the ecliptic cut at two points. The point corresponding to the Sun's apparent passage from south to north of the equator is known as the First Point of Aries, and is usually labelled γ.

Figure 4.1 shows these planes, points and their relative positions.

4.2   CO-ORDINATE SYSTEMS

The position of a celestial body projected on to the celestial sphere can be specified in several ways. Two systems are used more frequently than the others, while a third is often used because it is convenient on some occasions in spite of its disadvantages.

(a)   Celestial latitude and longitude

This is very similar to latitude and longitude as applied to the Earth's surface. The reference for latitude on Earth is the equator, but for the celestial sphere the reference for latitude is the ecliptic. All latitudes on the north side of the ecliptic are regarded as positive, and those on the south side negative. The latitude of the body in Figure 4.2 is marked β. The reference for longitude on Earth is the plane passing through the north pole and Greenwich. For the celestial sphere, the reference plane passes through the north pole of the ecliptic and the First Point of Aries (γ). Celestial longitude is marked λ in Figure 4.2. It is measured from $0°$ to $360°$ anti-clockwise looking at the ecliptic from the north pole of the ecliptic.

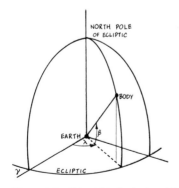

Figure 4.2   Celestial latitude λ and longitude β

(b)   Declination and right ascension

This system, Figure 4.3, is similar to latitude and longitude, but the celestial equator is taken as reference for declination. Declination (δ) is regarded as positive when the body is north of the celestial equator, and negative when south of the equator. Right ascension (a) is measured from the plane passing through the north celestial pole and the First Point of Aries. It is measured from 0° to 360° anti-clockwise looking at the equator from the north celestial pole.

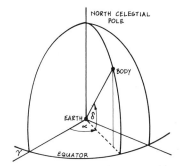

Figure 4.3   Right ascension a and declination δ

(c)   Altitude and azimuth

This system is convenient to use since the observer's horizon is taken as reference for altitude. The angle which the line joining the observer to the celestial body makes with the horizontal is known as the altitude of the body. The azimuth is the angle measured from the north point to the celestial body. It therefore ranges either from 0° to 180° to the east or from 0° to 180° to the west.

The altitude and azimuth system is much used by amateur astronomers, but it is not the best system to use when setting a telescope to follow an astronomical body. Both the altitude and the azimuth are changing with time, and a double adjustment of the telescope is necessary to keep the object in view.

If the axes of a telescope are arranged so that one of them points to the north celestial pole, it will only be necessary to rotate the telescope about this single axis to keep the celestial body in view. Such a mounting is referred to as an equatorial mounting. Most of the large telescopes of the world, such as the 200 inch Hale reflecting telescope at Mount Palomar, are equatorially mounted but the relatively recently installed Soviet 238 inch reflecting telescope, which is now the largest optical telescope, has an altitude-azimuth mounting.

4.3 ZENITH ANGLE

The angle which the line joining the observer to the celestial body makes with the line joining him to the zenith, is known as the zenith distance. It is, in fact, 90° − the altitude of the celestial object.

As we move on the surface of the spherical Earth, the direction of our zenith will change. Any relation, therefore, which connects altitude–azimuth co-ordinates with the corresponding equatorial co-ordinates will include the terrestrial latitude φ of the observer.

4.4 SIDEREAL TIME

In Section 5.12 we define the hour angle of the Sun as the angle which the great circle through the Sun and the celestial poles makes with the meridian. The hour angle of any other celestial body, or even any position in the sky where there is no body, is defined in the same way.

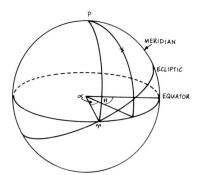

Figure 4.4   Sidereal time

The hour angle of the First Point of Aries γ (see Section 4.1) is thus defined as the angle between the great circle through γ and the celestial poles and the meridian. The hour angle of γ at any place is known as the sidereal time.

Referring now to Figure 4.4, we see that the hour angle of a star X is the angle H, and that the right ascension of star X is the angle a (see

Section 4.2). It is also clear that the hour angle of γ, that is the sidereal time, is the sum of H and a.

Thus,   sidereal time = right ascension of a star + hour angle of the star.

The sidereal time corresponding to the time in everyday common use is listed in such publications as the *Handbook of the British Astronomical Association.*

4.5   PROJECT 14

To construct a nomograph for the conversion of equatorial co-ordinates to altitude–azimuth co-ordinates, and vice versa.

The co-ordinates of planets, stars and other celestial bodies are most often given in official publications in equatorial co-ordinates, that is, right ascension a and declination δ.

It is probably easier for the amateur astronomer to find an object in the sky if he knows the alt–azimuth co-ordinates, that is, the altitude, a (or alternatively the zenith distance z) and the azimuth angle A (see Section 4.2). This means that an easy method of conversion is desirable. The nomograph in this project provides a quick graphical means of conversion of sufficient accuracy for many purposes. It is, of course, based on the mathematical relations between the co-ordinates which can be deduced by spherical trigonometry, (see Appendix A3).

Two nomographs are involved, the first connecting declination δ and hour angle H (and therefore right ascension by the formula in Section 4.4) with zenith angle z, and the second connecting declination δ and zenith angle z with azimuth A.

Both nomographs take the form shown in Figure 4.5 .

First set up $x-y$ axes in the normal way with the origin O at the left hand edge of the graph paper and mid-way between top and bottom.

Use a scale of   10 mm = 0.1 unit   throughout.

Taking angles of $0°$ to $180°$ in intervals of $10°$ for H plot the points $(0, -\cos\varphi\cos H)$ where φ is the terrestrial latitude of the observer. Label the H axis in time rather than angle so that $0°$ is marked 0 hours and 24 hours, $30°$ is marked 2 hours and 22 hours, $60°$ is marked 4 hours and 20 hours and so on. This is because the Earth rotates $360°$ in 24 hours or $15°$ per hour.

Taking angles of $0°$ to $90°$ in intervals of $10°$ for z plot the points $(1, \cos z)$.

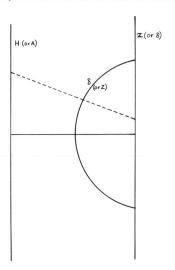

Figure 4.5   Co-ordinate nomograph

Taking angles of $-90°$ through zero to $+90°$ in intervals of $10°$ for δ plot the points

$$\left(\frac{1}{1+\cos \delta}, \ \sin \varphi \ \tan \frac{\delta}{2}\right).$$

From this nomograph we can determine the zenith angle z of a celestial object given its right ascension a and its declination δ at a given time. First, using the formula in Section 4.4

$$\text{sidereal time} = a+H$$

or $$H = \text{sidereal time}-a$$

we calculate the hour angle H. The sidereal time corresponding to the time in everyday common use – Greenwich Mean Time – is listed in such publications as the *Handbook of the British Astronomical Association*.

Draw a straight line from H just determined through the value of δ on the nomograph to cut the zenith axis in z the zenith angle required.

As an example a line from $H = 15$ hours   through $\delta = +50°$ should cut the zenith axis in $z = 71°.5$ or altitude $a = (90 - 71°.5) = 18°.5$ if the diagrams are drawn for latitude $\varphi = 51°.5\ N$.

The second nomograph relating z, δ and A, the azimuth, must now be drawn.

On a second sheet of graph paper set up $x - y$ axes as before and use the scale of $10 \, \text{mm} = 0.1 \, \text{unit}$ throughout.

Taking angles of $0°$ to $180°$ in intervals of $10°$ for A plot the points $(0, - \cos \varphi \cos A)$.

Taking angles of $- 90°$ through zero to $+ 90°$ in intervals of $10°$ for δ plot the points $(1, \sin \delta)$.

Taking angles of $0°$ to $90°$ in intervals of $10°$ for z plot the points

$$\left(\frac{1}{1 + \sin z}, \; \sin \varphi \, \frac{\cos z}{1 + \sin z} \right)$$

Continuing the example, if we draw a straight line from $\delta = + 50°$ through $z = 71.5°$ this should cut the azimuth axis in $A = 28°$.

Thus $(H = 15 \text{ hours}, \quad \delta = + 50°)$ converts to $(a = 18°.5, \quad A = 28°)$.

4.6   MAPS OF THE SKY

The Earth is approximately a sphere, and is often represented by a globe with the shapes of the countries on it. This is not a very convenient form for everday use, especially if much detail is required, since the globe would have to be uncomfortably large. Parts of the Earth's surface can be represented on flat pieces of paper in the form of maps, places being shown on a grid of latitude and longitude.

Similarly, although celestial globes are available, maps of part of the sky are more convenient. The grid on these maps may be celestial latitude and longitude, or more likely, declination and right ascension. Right ascension is often converted from degrees to time measure. $360°$ are equivalent to 24 hours, or $1°$ degree is equivalent to 15 minutes of time, so that R.A. may be quoted in hours, minutes and seconds.

Such a map showing part of the sky is demonstrated in Figure 4.6. The stars fall into groups called constellations. Stars of any one constellation are not necessarily near to one another, but appear to be so because the eye cannot judge the relative distances to them. The constellations have names associated with likenesses which people in the past have seen to animals, men and women and their gods. There are some very beautiful star maps with these likenesses drawn or painted in, but today the divisions between stars of different constellations are drawn in on star maps

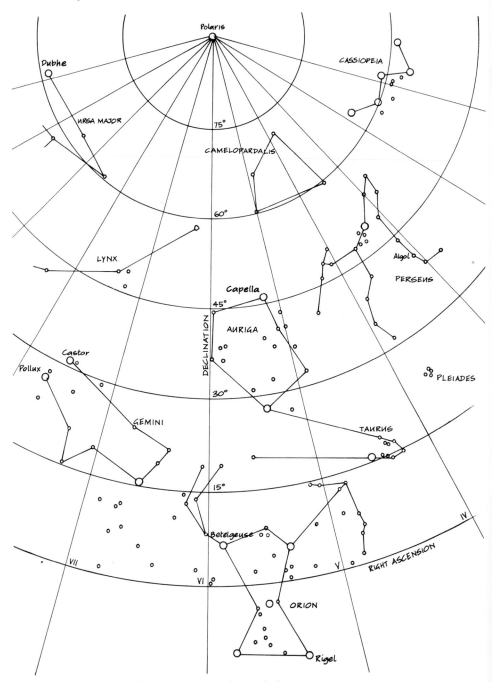

Figure 4.6   General star map and constellations

following the grid of declination and right ascension or lines parallel to these.

It is clear that the curved surface of a globe cannot be represented on a flat piece of paper without some distortion. There are many geometrical constructions, called map projections, each of which will keep some special feature on the flat paper as on the globe. Thus we may have equal area projection in which areas are translated faithfully at the expense of distortion at high latitudes. Mercator's projection has the feature that a uniform compass course is represented by a straight line, but again there is distortion, in that countries in high latitudes are shown to a much larger scale than those near the equator.

One such projection which lends itself to astronomical maps, provided the spread in right ascension is not too great, is the polyconic projection. This is an improvement on the simple conical projection, shown in Figure 4.7.

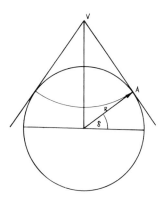

Figure 4.7   Conical projection

4.7   POLYCONIC MAP PROJECTION

Consider a cone, vertex V, sitting on the globe and touching it at declination δ (Figure 4.7).

The circumference of the circle of declination δ is $2\pi R \cos \delta$, where R is the radius of the globe. Thus when the cone is opened out into the sector of a circle, the circle of declination will be represented by the arc of a circle with centre V and of radius VA, the length of the arc being $2\pi R \cos \delta$, as shown in Figure 4.8.

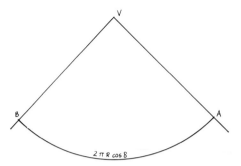

Figure 4.8   Development of cone

In polyconic projection, there is a different cone for each circle of declination. If these are drawn on a single piece of paper which may be thought of as rolling up the sphere, the distance between the lines (circles) of declination at the centre of the map will be the same as the curved distance between the corresponding adjacent declination circles on the globe. Each circle on the flat paper will have a radius equal to the corresponding VA, and a length equal to its own $2\pi R \cos \delta$. The centres of these circles will, however, move downwards from V as the declination increases. Thus a grid of declination and right ascension will be set out as in Figure 4.9.

In this projection, the scale along the parallels of declination is true, as is the scale along the central meridian of the map. As the right ascension increases from the centre, the lines of declination get wider apart, and this limits the size of the map in right ascension.

4.8   PROJECT 15

To construct a map of a small portion of the sky.

The map to be constructed in this project covers that portion of the sky ranging from 45° on one side of the meridian to 45° on the other side of the meridian for right ascension, and 60° on each side of the equator for declination.

The radius of the globe from which the map is to be constructed is 100 mm. This means that the radius of the largest circle to be drawn is a little under 600 mm. Compasses do not normally stretch to this radius, and once the radii have been calculated, it will be found profitable to make a beam compass from a strip of wood, a nail and a pencil.

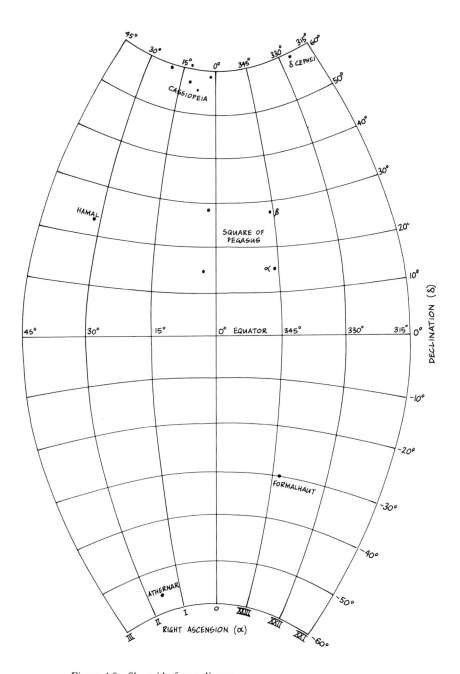

Figure 4.9   Sky grid of co-ordinates

The construction of this device is indicated in Figure 4.10. The position of the nail is marked on the strip of wood near to one end. Each radius required is marked from the nail as centre, and a hole drilled just large enough to hold the pencil firmly. Alternatively string may be used, but it is a poor substitute for the simply constructed beam compass.

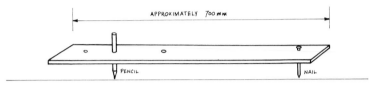

Figure 4.10   Beam compass

Calculate the radius of each of the circles of declination from the formula

$$z = \frac{R}{\tan \delta}$$

where R is the radius of the globe   (= 100 mm)   and δ is the declination, for values of δ from 10° to 60° inclusive, in steps of 10°.

Calculate the distance on the circumference of the globe between the parallels of declination 10° apart, using the expression $\pi R/18$. Draw a horizontal line centrally on a piece of paper to represent the equator, and a second, vertical, line centrally on the paper to represent the meridian. Let these lines intersect in O.

From O, measure six intervals on each side of the equator spaced $\pi R/18$ apart. Let the one for declination 10° cut the meridian in A. From A, measure a distance of $z = 100/\tan 10°$ along the meridian away from O. Let this point be B. With centre B and radius z, draw an arc with A as its middle point. Then this is the circle for declination 10°.

Repeat for the second interval along the meridian (20°) and radius $z = 100/\tan 20°$, and so on for the six intervals on one side of the equator, and then for the six intervals on the other side of the equator.

The circumference of each circle of declination will be $R \cos \delta \times 2\pi$ and the length of the arc between circles of right ascension spaced 15° (= 1 hour) apart will be one twenty-fourth part of this. Calculate this quantity for each value of δ. Starting at the meridian, mark three intervals of right ascension on each side of the meridian, each interval being the value last calculated for that declination.

Join with a smooth curve corresponding points for right ascension 15°, then those for 30° and so on. These curves form the circles of right ascension, and the whole grid forms the sky map frame (Figure 4.9) on to which can now be put the stars. Some stars in this portion of the sky are listed below with their co-ordinates so that they can be marked on the map.

TABLE 4.1

Star	Declination	Right ascension
Formalhaut	−29°56′	22h 55m
Hamal	+23 19	2 06
Achernar	−57 23	1 36
δ Cephei	+58 16	22 28

4.9 PROJECT 16

(i)   To plot the path of a comet on a star map.
(ii) To plot the path of the Sun on the same star map.
(iii) To determine the angle of obliquity of the ecliptic.

Data

The Table below gives the declination and right ascension of Comet Arend Roland and of the Sun for similar dates. The values for the comet were computed from a few observations, and gave its predicted path in the sky.

Using the grid on the star map constructed in Project 15, plot the positions of the comet for the dates given in the data table, and sketch the path of the comet. Label each point with the date.

Similarly, using the co-ordinates of the Sun given in the data table, plot its position on the star map, and label each point with the date. Draw the path of the Sun. This will be the ecliptic.

Since the scales of declination and right ascension are correct on the central meridian when using polyconic projection, the angle made by the path of the Sun at zero declination and zero right ascension with the equator will also be true to scale. Measure this angle, and compare it with the angle of obliquity of the ecliptic which at that time was 23°27′.

TABLE 4.2

Date		Comet Arend Roland		Sun	
		δ	α	δ	α
1957 Mar	18			−0°56′	23h 51m
	22			+0 38	0 06
	23	−11°09′	0h 35.6m		
	25	−11 41	0 36.4		
	26			+2 13	0 20
	30			+3 46	0 35
	31	−12 58	0 38.7		
Apr	3			+5 19	0 50
	6	−12 02	0 42.1		
	7			+6 50	1 04
	8	−10 24	0 44.5		
	10	−7 39	0 48.3		
	11			+8 19	1 19
	12	−3 35	0 54.1		
	14	+1 54	1 02.1		
	15			+9 46	1 34
	16	+8 42	1 12.8		
	18	+16 32	1 26.2		
	19			+11 11	1 48
	20	+24 48	1 42.5		
	22	+32 51	2 01.5		
	23			+12 32	2 03
	24	+40 07	2 23.2		
	26	+46 16	2 47.1		
	27			+13 50	2 18
	28	+51 12	3 12.7		

Although Comet Arend Roland was easily visible to the naked eye, this project shows that it would be of no use to look for it around the dates April 12 to April 18, since the glare of the Sun would prevent it from being seen. A record exists that the comet was seen on April 20.

The photograph (Figure 4.11) shows the comet as it was seen on 1957 April 25 when it was well removed from the Sun.

The great majority of comets cannot be seen with the naked eye, and their location is therefore much more difficult. Forecasts of their paths

Figure 4.11   Comet Arend Roland   *Photo © Armagh Observatory, College Hill,*
Armagh, Northern Ireland.
Reproduced by kind permission

can be computed from a few observations of the comets, and this project is useful in determining when not to search for a particular comet.

4.10 THE ZODIAC

As Project 16 shows, the Sun appears to move back along the ecliptic relative to the stars about one degree per day, so that in just over 365 days it has returned to the same position in the sky. Its path takes it through a number of constellations, or groups of stars, which form a rough belt across the sky called the zodiac.

These constellations are:

Aries	–	the ram
Taurus	–	the bull
Gemini	–	the twins
Cancer	–	the crab
Leo	–	the lion
Virgo	–	the virgin
Libra	–	the scales
Scorpio	–	the scorpion
Sagittarius	–	the archer
Capricornus	–	the he-goat
Aquarius	–	the water bearer
Pisces	–	the fishes

The position of the Sun is often located approximately by saying that it is in a particular constellation of the zodiac. Thus, the statement that the Sun is "in Gemini" means that the Sun is in the constellation of Gemini at that time. Astrologers believed, and some still do, that the position of the Sun in a particular constellation of the zodiac when a person was born, governed the character and life of that person. Nowadays, horoscopes are often issued by the sign of the zodiac according to the date of one's birth in relation to the limits set for each 'sign'. The spread of each constellation of the zodiac for this purpose is equal to the others, that is, each occupies 30° of the ecliptic.

The boundaries of the constellations on star maps were drawn arbitrarily relative to the stars by Monsieur Delporte, and adopted by the International Astronomical Union in 1930. For convenience the boundaries are parallel to circles of declination and right ascension. Due to precession of the Earth's axis, the ecliptic has changed its position relative to the constellations, and it is interesting to see the discrepancies which exist between the astronomer's definition that the Sun is in a particular constellation and the dates set by the astrologers. In fact, it will be noticed that the Sun passes through fourteen constellations in one revolution of the ecliptic, the extra constellations being Ophiuchus, the serpent bearer and, briefly, Cetus, the whale.

4.11   PROJECT 17

To determine the dates when the Sun enters and leaves various constellations in the zodiac, and to compare these dates as specified by astrologers.

Data

The position of the Sun throughout 1970 is given by the co-ordinates below.

TABLE 4.3

Date		Declination δ	Right ascension a
Jan	1	−23°01′	18h 46m
	9	−22 07	19 21
	17	−20 46	19 56
	25	−19 00	20 30
Feb	2	−16 51	21 03
	10	−14 23	21 35
	18	−11 41	22 06
	26	−8 46	22 37
Mar	6	−5 43	23 07
	14	−2 35	23 36
	22	+0 35	0 05
	30	+3 43	0 34
Apr	7	+6 47	1 04
	15	+9 43	1 33
	23	+12 29	2 03
May	1	+15 02	2 33
	9	+17 19	3 04
	17	+19 18	3 54
	25	+20 56	4 07

TABLE 4.3 *continued*

Date		Declination δ	Right ascension α
Jun	2	+22°10′	4h 40m
	18	+23 24	5 46
	26	+23 22	6 19
Jul	4	+22 54	6 53
	12	+22 00	7 25
	20	+20 42	7 58
	28	+19 01	8 29
Aug	5	+17 01	9 00
	13	+14 43	9 31
	21	+12 10	10 01
	29	+9 25	10 30
Sep	6	+6 30	10 59
	14	+3 28	11 28
	22	+0 22	11 57
	30	−2 45	12 25
Oct	8	−5 50	12 54
	16	−8 50	13 24
	24	−11 42	13 54
Nov	1	−14 23	14 25
	9	−16 49	14 57
	17	−18 57	15 29
	25	−20 43	16 03
Dec	3	−22 05	16 37
	11	−22 59	17 12
	19	−23 25	17 48
	27	−23 20	18 23

The Table below gives the dates which astrologers use for the various constellations.

The maps in Figure 4.12(a–d) show the sky in the region of the zodiac with the boundaries of the constellations represented by dotted lines.

Plot on the map the position of the Sun at dates near the boundaries of the constellations. Estimate the date when the Sun enters and the date when the Sun leaves each constellation. Compare these dates with the dates for each constellation as used in horoscopes.

TABLE 4.4

Aries	March	21 – April	20
Taurus	April	21 – May	20
Gemini	May	21 – June	20
Cancer	June	21 – July	20
Leo	July	21 – August	21
Virgo	August	22 – September	22
Libra	September	23 – October	22
Scorpio	October	23 – November	22
Sagittarius	November	23 – December	20
Capricornus	December	21 – January	19
Aquarius	January	20 – February	18
Pisces	February	19 – March	20

4.12   METEORS

We have most of us seen, usually out of the corner of the eye, a very quick streak of light in the night sky as if a star has moved its position rapidly. We may even have seen the trail of light persist for a very short time.

What we have seen is most probably a meteor entering the Earth's atmosphere. Meteors are, on the whole, very small particles about the size of a grain of sand. They travel at speeds of about 30 km/s, although their velocities can be much higher than this.

Thus the friction between the meteor and the Earth's atmosphere causes the incandescence which we see, and the burning up of the meteor before it reaches the surface of the Earth. A very small percentage of the meteors do not burn up completely, and these reach the surface of the Earth. Many remains of meteors have been found. The term 'meteorite' is applied to the latter.

Meteors are thought to be space debris left behind by comets, possibly sometimes by the breaking up of a comet. The meteors then tend to be spread all along the orbit of the comet, and if the orbit of the Earth intersects this orbit, we get a large number of meteor trails in a short time. This is referred to as a shower of meteors. The Earth, in its orbit, will cut the meteor orbit at about the same time each year, and we are able to forecast these showers.

Observation of these showers brings out the fact that the lines of meteor

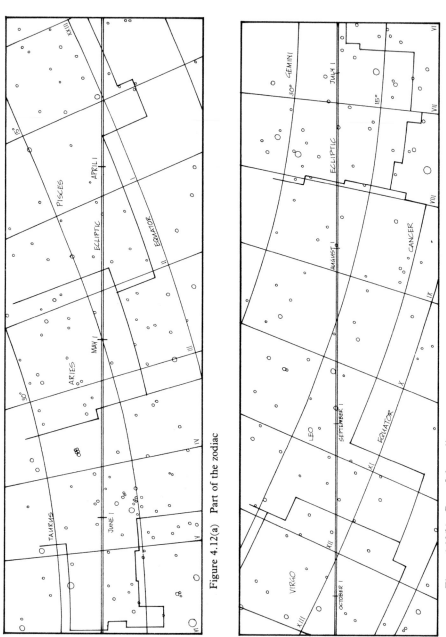

Figure 4.12(a)   Part of the zodiac

Figure 4.12(b)   Part of the zodiac

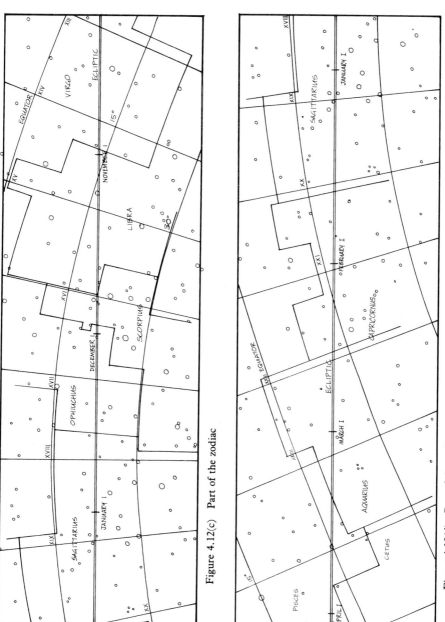

Figure 4.12(c)   Part of the zodiac

Figure 4.12(d)   Part of the zodiac

trails when produced, appear to diverge from a single point in the sky known as the radiant. The particular shower observed is known by the name of the constellation in which the radiant appears to be. Thus we have the Lyrids (constellation Lyra), the Perseids (constellation Perseus), the Andromedids (constellation Andromeda) and so on.

The divergence of meteor trails from a radiant is really the result of perspective observation. Since the meteors in any one shower are travelling in an orbit, they must be travelling in parallel paths.

If we first consider all those meteors which are travelling in parallel paths in a plane which also passes through the observer, Figure 4.13 makes it clear that we see the projections of these paths on the celestial sphere, and that these projected paths all seem to come from the point R. The observed paths are all parallel to the line joining the observer to R, that is, parallel to OR.

Figure 4.13   Meteor trails and radiant

It follows that whatever plane we choose, the meteors on it will have a motion parallel to OR, and the same reasoning can be applied to these with regard to the radiant R. Thus to the observer, the shower of meteors will appear to radiate from the radiant R.

4.13   PROJECT 18

To plot on a map of the sky the tracks of a number of meteors of the Perseid shower, as actually observed on the night of 1967 August 11/12, and to determine the equatorial co-ordinates of the radiant of the meteor stream.

TABLE 4.5

Start		End	
Right ascension	Declination	Right ascension	Declination
1h 00m	50.5°	23h 45m	32.0°
0 51	45.0	0 24	38.5
1 42	44.0	1 21	32.0
2 18	52.0	1 33	33.5
3 03	44.0	3 03	35.5
3 21	48.5	3 39	30.0
4 03	42.5	4 18	32.0
4 18	56.0	6 12	42.0
4 51	66.5	6 39	64.0
5 21	77.0	10 45	76.0
2 00	81.0	17 30	81.0
23 39	76.0	19 33	67.0
22 48	64.0	20 18	48.0
23 48	62.5	21 33	51.0
1 21	60.0	22 54	46.5

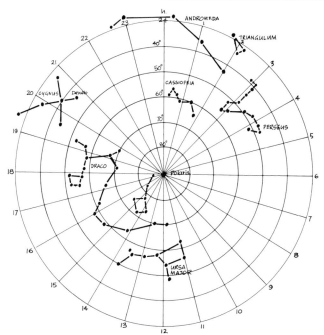

Figure 4.14   Map for Perseid meteor shower

Reproduce the map of the sky as shown in Figure 4.14 using a scale of declination (radially from the centre of the map) of 10 mm = 10°, and dividing the full circle of 24 hours of right ascension into 15° = 1 hour.

From the data table, plot on the map the start and end of the first meteor trail. Join these points with a straight line. Repeat for all the other meteor trails.

Faintly produce each of these trails in a backwards direction. They will be seen to converge into a small area in the constellation of Perseus. Locate the approximate centre of this area, and mark it *R*. This is the apparent radiant.

Read off the co-ordinates of *R*, and compare these with right ascension 3h (45°) and declination + 58°.

4.14   DIRECT AND RETROGRADE MOTION OF THE PLANETS

Observation of those planets like Mars and Jupiter whose orbits lie outside the orbit of the Earth relative to the Sun, shows that, in general, they move eastwards relative to the background of stars as time progresses. Since Jupiter is further away its motion against the background of stars is much less than that of Mars, but in both cases the motion is referred to as direct motion. This motion must not be confused with the daily westward motion, which is due to the rotation of the Earth on its axis.

At times, however, a contradiction of the above takes place, and one or other of these planets appears to move westwards relative to the stars. This motion, known as retrograde motion, proceeds for only a relatively short time before direct motion is resumed. During the retrograde motion and the following direct motion, the path of the planet may be a loop in the sky. Before the recognition that the planets travel round the Sun and that the Earth is not the centre of the solar system, explanation of the above motions proved to be an embarrassment to those who maintained that our planet was indeed the centre of the universe.

Project 19 gives details of the co-ordinates of Mars as seen from the Earth during such a period of direct and retrograde motion. The motion of Mars can be put on to a sky map and the movement observed for this period.

Project 20 is designed to illustrate how this motion is derived simply from the relative positions of Mars and the Earth at certain times in their orbits round the Sun. The inclination of the plane of the orbit of Mars to that of the Earth is only 1.8 degrees. For this purpose it is assumed that the orbits are coplanar. Also, the eccentricities of both orbits are so small that to assume the orbits are both circular will not materially affect the explanation.

4.15 PROJECT 19

To plot on a sky map the apparent path of Mars as seen from the Earth when retrograde motion is taking place.

Data

The Table below gives the declination and right ascension of Mars for a period in 1967.

TABLE 4.6

Date		Right ascension	Declination
Jan	10	13h 10.4m	−5°09′
	20	13 26.0	−6 38
	30	13 39.9	−7 53
Feb	9	13 51.6	−8 54
	19	14 00.6	−9 38
Mar	1	14 06.1	−10 04
	11	14 07.3	−10 09
	21	14 03.8	−9 52
	31	13 55.4	−9 14
Apr	10	13 42.8	−8 18
	20	13 28.3	−7 14
	30	13 14.3	−6 17
May	10	13 03.4	−5 39
	20	12 57.1	−5 27
	30	12 55.8	−5 45
Jun	9	12 59.2	−6 28
	19	13 06.8	−7 33
	29	13 17.8	−8 56
Jul	9	13 31.8	−10 32

The area of sky involved in this exercise is so small that it is not necessary to get involved in map projection.

Set up a grid of lines ranging from right ascension 13 hours on the right to 14 hours on the left, and ranging from declination − 5° at the top to − 15° at the bottom. Scales of 10 mm = 5 minutes of time in R.A. and 10 mm = 1° in declination will be found suitable.

Using the data table, plot the position of Mars for each of the dates given, and join up the points with a smooth curve.

Observe that Mars has direct motion (to the left) at first, and then retrograde motion (to the right), and finally resumes its direct motion.

The turning points round about March 11 and May 30 are known as stationary points, as the planet is moving neither directly nor retrograding at these dates.

4.16   PROJECT 20

(i)   To plot the positions of the Earth and Mars in their orbits round the Sun for corresponding dates.

(ii) To show how the line of sight from the Earth to Mars alters in such a way as to give the impression that Mars is retrograding.

Data

The Table below gives the heliocentric longitude of the two planets for various dates in 1967.

With the paper longways on, and centre to the right hand side of the paper, draw a circle of radius 25 mm to represent the orbit of the Earth with the Sun as centre.

The radius of the orbit of Mars is a little over 1.5 times that of the Earth, and so a circle with the same centre but of radius 37.5 mm will represent this orbit.

From the centre of the circles draw a radius horizontally to the right. Take this as reference for the angles of longitude.

For the Earth, measure the angles of longitude anti-clockwise from the reference line, and mark each position on the Earth circle 1, 2, 3 etc., in order from the top of the data table.

TABLE 4.7

Date		Heliocentric longitude (°)	
		Mars	*Earth*
Jan	10	161.7	108.8
	30	170.5	129.2
Feb	19	179.3	149.4
Mar	11	188.3	169.5
	21	192.8	179.5
	31	197.4	189.4
Apr	10	202.0	199.2
	20	206.7	209.0
	30	211.4	218.7
May	10	216.2	228.4
	20	221.1	238.1
	30	226.1	247.7
Jun	19	236.2	266.8
Jul	9	246.7	285.9

For Mars do likewise.

Draw a line at right angles to the reference line 150 mm from the centre of the circles. Join 1 on the Earth circle to 1 on the Mars circle, and produce to cut the last line drawn. Join 2 to 2, 3 to 3 and so on. Then the projections of these lines trace out the path of Mars as it would be on the celestial sphere. Join them up in order of number, slightly displacing the direct and retrograde motions so that these lines are not drawn on top of one another.

Determine the stationary points, and compare their dates with the actual dates of March 11 and May 30.

4.17   THE INFERIOR PLANETS

The orbits of the planets Venus and Mercury lie inside the orbit of the Earth round the Sun. Mercury and Venus are referred to as the inferior planets.

Unlike the planets which have their orbits outside that of the Earth, the inferior planets do not exhibit continuous direct motion against the background of stars. They appear to oscillate backwards and forwards, with the Sun as the centre of the oscillation, between limits. Thus Venus appears to travel from a position in the sky such that the line joining it to the Sun subtends an angle of 46° at the Earth to a position on the other side of the Sun subtending the same angle. These extreme positions are known as the greatest elongation of Venus, east if the planet appears to be on the left of the Sun, and west if on the right.

When Venus is on the left of the Sun it is said to be an evening 'star' since it is visible in the night sky after the Sun has set (and sometimes, because of its great brilliance, before the Sun has actually set). Similarly, when Venus is on the right of the Sun it is visible in the morning sky before the Sun rises, and is therefore a morning 'star'.

In its oscillating motion, Venus appears to travel in front of the Sun, although it cannot be seen at this time because of the brightness of the Sun. The Earth, Venus and the Sun are then almost in line, except for the slight differences in the inclination of the planets' orbits, and inferior conjunction is said to occur. Similarly, when the planets and the Sun are in line again, but with Venus behind the Sun, superior conjunction is said to occur.

Project 21 shows how the oscillating motion of Venus takes place, how we can arrive at an approximate value for the greatest elongation, and how to determine the approximate dates for the two types of conjunction.

4.18   PROJECT 21

(i)   To plot the positions of the Earth and Venus in their orbits round the Sun at various corresponding dates.
(ii)   To determine the dates for inferior and superior conjunction.
(iii)   To determine the greatest elongation of Venus, and the dates on which it occurs.
(iv)   To plot the oscillating motion of Venus.

Data

The Table below gives the heliocentric longitudes of the Earth and Venus for corresponding dates throughout 1970 and the beginning of 1971.

The angle between the plane of the orbit of the Earth and the plane of the orbit of Venus is very small, and for the purposes of this project we

TABLE 4.8

Date			Heliocentric longitude (°)	
			Earth	Venus
1970 Jan		14	113.1	287.0
		24	123.3	302.8
	Feb	23	153.7	350.4
	Mar	15	173.7	22.2
	Apr	4	193.5	54.2
		24	213.1	86.4
	May	14	232.5	118.8
	Jun	3	251.1	151.3
		23	270.8	183.7
	Jul	13	289.9	216.0
	Sep	1	337.9	295.4
	Oct	1	7.1	342.9
		31	36.9	30.7
	Nov	10	46.9	46.7
		30	67.0	78.8
1971 Jan		9	107.8	143.7
		29	128.1	176.1
	Feb	18	148.4	208.4

can assume the orbits to be coplanar. Also the eccentricity of each orbit is small, so that we can assume the orbits to be circular.

With centre S, the Sun, draw a circle 50 mm radius to represent the orbit of the Earth. Since the radius of the orbit of Venus is 0.72 times that of the Earth, a circle with centre S and radius 36 mm can be drawn to represent the orbit of Venus.

From S draw a horizontal line to give a reference for measuring the longitudes of the planets. Using the data given in the Table, mark off the angles of longitude on the Earth circle, and label these positions of the Earth 1, 2, 3 etc. Do likewise on the Venus circle, marking the corresponding points 1, 2, 3 etc. Join point 1 on the Earth circle to point 1 on the Venus circle, and so on.

Note especially those points which, when joined, pass through S. The dates corresponding to these positions will be those of inferior and superior conjunction.

Note also those points which, when joined are tangential to the Venus circle. These are the dates of greatest elongation.

Compare the above results with:

Superior conjunction	1970 Jan 4
Inferior conjunction	1970 Nov 10
Greatest elongation east	1970 Sep 1
Greatest elongation west	1971 Jan 20

Finally, measure the perpendicular distances from S to each of the lines joining the Earth and Venus points. Draw a horizontal line, mark the centre S, and mark off from S these perpendicular distances to the left and right according to the position of the perpendiculars with respect to the Sun. Number them according to the successive positions of Venus as seen from the Earth. Note that the position of Venus as seen from the Earth oscillates along this line.

4.19   PROJECT 22

To make a magnetic board suitable for displaying (a) the positions of the Sun, Moon and planets and (b) the path of any visible artificial satellites.

(a)   The Sun moves about 30° along the ecliptic each month. The Moon also moves near the ecliptic about 13° every 24 hours and goes through its various phases in about 28 days. The planets, particularly the outer planets, do not appear to move very much in one month, but their positions are always near the ecliptic.

We shall therefore reproduce a map of the sky showing the ecliptic and the sky on each side of it. We shall mount the map on thin sheet steel, and use coloured magnetic discs, representing the heavenly bodies, which will cling to the map on the steel sheet to mark their current positions. They can easily be lifted off and replaced correctly for the succeeding month's display.

A most attractive map, Figure 4.15, can be drawn in white ink on thin black card about 255 mm × 205 mm. Choose a card which is non-absorbent and has a slight glaze on it so that the ink will not run. The monthly positions of the Sun, Moon and planets may be obtained from a publication such as the *Handbook of the British Astronomical Association*.

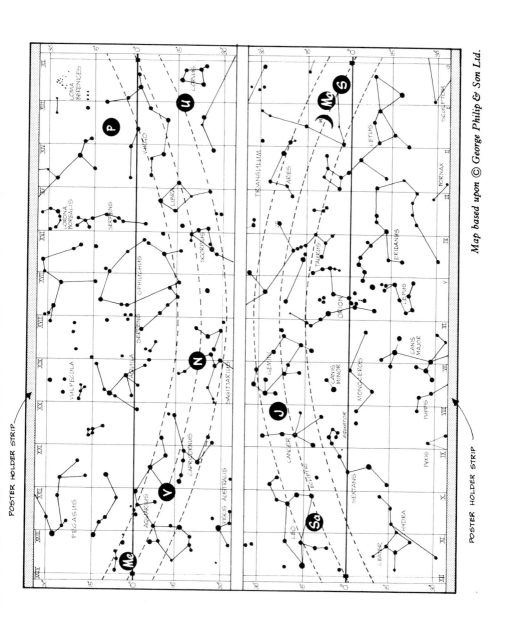

Figure 4.15 Magnetic display board (front)

Map based upon © George Philip & Son Ltd.

(b) On a second piece of black card the same size as above, copy Figure 4.16, the vertical lines representing the points of the compass and the horizontal lines altitudes. Mount this on the reverse side of the steel sheet above, or separately if desired. A thin strip of magnetic tape can then be positioned daily to represent the salient features of the path of any artificial satellite which may be visible. *The Guardian* gives such information daily on the back page under "Satellite Predictions" in the form Pageos 1, 22.00 – 22.18, ENE 15NNE N. These are respectively the time of visibility, where rising, maximum altitude and direction, where setting.

The top and bottom of each magnetic display board can be finished off neatly using plastic strip similar to that which can be bought for suspending wall posters.

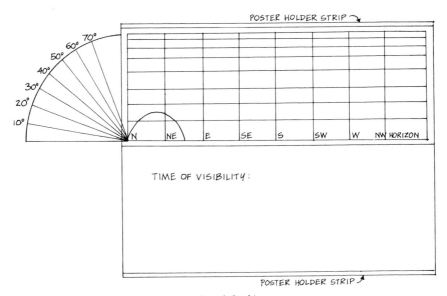

Figure 4.16   Magnetic display board (back)

5

The Sun

5.1 ECLIPSES

The spectacle of an eclipse is something most of us have seen. An eclipse occurs when the Earth comes directly, or almost directly, in between the Sun and the Moon, so that the shadow cast by the Earth in the Sun's rays falls on the Moon. The Moon appears to be wholly, or partially blotted out, and we say that this is a full or a partial eclipse of the Moon. Because of their sizes and their distances from the Sun, the diameter of the shadow cast by the Earth is larger than the diameter of the Moon, and a lunar eclipse can be seen over a large portion of the Earth.

This is not so when the Moon comes directly, or nearly directly, between the Sun and the Earth, and an eclipse of the Sun takes place. The shadow cast by the Moon on the surface of the Earth is much smaller than the Earth, and an eclipse of the Sun is observed only by those people in the shadow. If the whole of the Sun appears to be blotted out, we say that there is a total eclipse; if only part of the Sun is blotted out, a partial eclipse; and if the central portion only is blotted out during the central part of the eclipse, we say that the eclipse is annular.

It is much more likely, therefore, that we have seen an eclipse of the Moon rather than an eclipse of the Sun.

If we refer to Figure 5.1 we see that in eclipses there is a region, called the umbra, which theoretically is in total darkness since no light can reach it from the Sun. This is the region behind the Earth, on the Moon side, within the cone formed by the external tangents touching the Sun and the Earth.

Again referring to Figure 5.1, we see that there is another region, known as the penumbra, where some light penetrates. This region is formed by

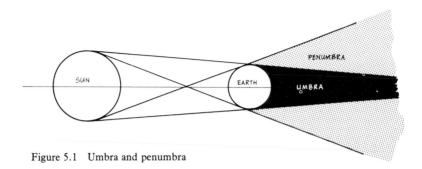

Figure 5.1   Umbra and penumbra

the internal tangents to the Sun and the Earth. It sometimes happens that the Moon passes through the penumbra without passing through the umbra. In these occurrences the Moon shows only a dimming, and is not blotted out.

5.2   EQUATORIAL PARALLAX

Astronomical distances are sometimes specified by the angle subtended at the distant object by a given baseline at the observer. Clearly, the more distant the object, the smaller will be this angle which is called the parallax of the object.

For stars which are very distant, it is necessary to have as large a baseline as possible, and the one taken is the radius of the Earth's orbit. For nearer objects, such as the Sun or the Moon, we can use the radius of the Earth itself as our baseline. The angle subtended at the Sun by the radius of the Earth is known as the equatorial parallax of the Sun, and similarly for the Moon. Since the Earth's orbit round the Sun is not truly circular, the equatorial parallax of the Sun will vary slightly throughout the year, and for a similar reason so will the equatorial parallax of the Moon throughout a lunar month. The values of these parallaxes at any time can be calculated, and, as will be shown in the following section, they can be used to find the size of the Earth's shadow through which the Moon passes in an eclipse.

5.3   ANGULAR RADIUS OF THE EARTH'S SHADOW

The size of the Earth's shadow will depend upon where it is measured. Since we are concerned with eclipses of the Moon, we need to find the radius of the Earth's shadow at the distance of the Moon, that is, in the Moon's orbit. In Figure 5.2 the actual radius of this shadow is MN, and the angular radius is the angle s. The angular radius of the Sun is s_r, the equatorial parallaxes of the Sun and Moon are P_s and P_m respectively, and we shall also need to use the semi-angle of the shadow cone which is marked v.

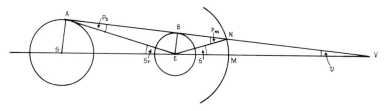

Figure 5.2   Radius of Earth's shadow

We shall use the property of triangles which gives the exterior angle of any triangle as the sum of the two interior opposite angles.

Thus, in triangle AEV,

$$s_r = P_s + v$$

Also, in the triangle ENV,

$$P_m = s + v$$

Subtracting

$$s_r - P_m = P_s - s$$

Hence, the angular radius of the Earth's shadow at the moon is given by

$$s = P_s + P_m - s_r$$

5.4   PROJECT 23

(i)   To plot the relative positions of the Earth's shadow and the Moon during an eclipse of the Moon.

(ii) To deduce the time and position of first contact and of last contact of the Moon with the Earth's shadow.

(iii) To determine the magnitude of the eclipse, that is, the proportion of the Moon's diameter which passes into the umbra.

Data

The data given below refer to the partial eclipse of the Moon which took place on 1970 August 17.

TABLE 5.1

Moon

Time	Right ascension	Declination
Aug 17d 00h	21h 37m 49s	−15°13′31″
01	21 40 13	−14 57 53
02	21 42 37	−14 42 09
03	21 45 01	−14 26 20
04	21 47 24	−14 10 25

TABLE 5.2

Sun

Time	Right ascension	Declination
Aug 17d 00h	9h 44m 01s	+13°37′48″
01	9 44 10	+13 37 01
02	9 44 20	+13 36 13
03	9 44 29	+13 35 26
04	9 44 38	+13 34 39

Equatorial horizontal parallax of the Moon	= 61′ 24″
Equatorial horizontal parallax of the Sun	= 09″
Radius of Sun	= 15′ 48″
Radius of Moon	= 16′ 44″

In this project we shall imagine ourselves to be located at the Sun, and watch the eclipse as if through a transparent Earth. In this way we shall see an apparently stationary Earth shadow being approached by the Moon.

The right ascension (see Section 4.2) of the Earth's shadow will be 12 hours removed from that of the Sun, since they are diametrically opposed. The declination (see Section 4.2) of the Earth's shadow will similarly have the same value as the declination of the Sun, but will be opposite in sign.

With these comments in mind, calculate for each of the times given above the quantities $(a_m - a_{es})$ and $(\delta_m - \delta_{es})$ where the suffix es refers to the Earth's shadow. The difference of right ascension will be in minutes and seconds of time. Convert these to seconds of arc by multiplying each of the values by $15'$ cos δ_m for the time involved. For an explanation of the cos δ_m factor see Section 8.4.

Next set up axes of δ vertically and a cos δm each to a scale of 10 mm $= 10'$.

Calculate the radius of the Earth's shadow from the parallaxes of the Sun and the Moon, and the solar radius. With centre the origin, draw the Earth's shadow to the same scale as those for the axes. Using the differences of right ascension and of declination, plot the relative position of the centre of the Moon for each of the times indicated in the data. These should fall on an approximately straight line which will be the path of the Moon's centre relative to the Earth.

With centre the origin, and radius equal to the sum of the radius of the Earth's shadow and the radius of the Moon, scribe the two arcs which cut the Moon's path. Then these will give the positions of the Moon's centre at first and last contact with the Earth's shadow. With each of these centres, and radius equal to that of the Moon, draw in the Moon's outlines.

Join the centre of the Earth's shadow to the centre of the Moon at first contact and so obtain the position of the Moon, as the angle from the north point, at which first contact is made. Do this similarly for last contact, and compare with the values 14° E and 63° W respectively.

Find the mid-point of the Moon's centres for first and last contact, and so draw the outline of the Moon at the maximum phase of the eclipse. Join the Moon's centre to that of the Earth's shadow. Measure that portion of the Moon in shadow, and express it as a fraction of the Moon's diameter. Compare this with the value of the magnitude of the eclipse, which was 0.413.

Finally, from the positions of the Moon's centre at the specified times, make an estimate of the times of first and last contact using the positions of the Moon's centre at contact. Compare these with Aug 17d 02h 18m and Aug 17d 04h 30m respectively. The time of central eclipse on this occasion was Aug 17d 03h 24m.

5.5   THE EQUATION OF TIME

To an observer on Earth, the Sun appears to move in an elliptical orbit with the Earth at one focus. It completes an orbit in what we call one year, and it might be thought that the position of the Sun in its orbit could be used for marking time. This is so, but it proves to be a not very convenient method.

The radius vector from the Earth to the Sun sweeps out equal areas in equal times in accordance with Kepler's second law of planetary motion. The radius vector changes in length, and cannot therefore sweep out equal angles in equal times, a feature necessary for the uniform marking of time. In other words, the apparent motion of the Sun along the ecliptic is not uniform with time.

Events on Earth are closely related to the movement of the Sun, and it is important that a system of time-keeping should be linked with the Sun.

A fictitious body, known as the dynamical mean Sun, is defined, which moves in the ecliptic, but, unlike the Sun it moves at a uniform rate. Thus the radius vector from it to the Earth sweeps out equal angles in equal times, completing one revolution around the Earth in one year. Its mean daily motion, as it is called, is

$$n = \frac{360}{365\frac{1}{4}} \text{ degrees per day}$$

A special feature of the dynamical mean Sun is that it is made to co-incide with the actual Sun when it is at perigee, round about January 1 (see Figure 5.3).

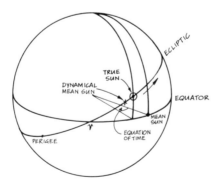

Figure 5.3   Equation of time

This arrangement is still not very satisfactory in that the Earth's axis is perpendicular to the equator and not to the ecliptic. The combination of a Sun moving in the ecliptic, be it the true Sun or the dynamical mean Sun, and the daily rotation of the Earth about the axis of the equator leads again to a non-uniform system of time.

A second hypothetical Sun, known as the mean Sun (see Figure 5.3 again), is therefore defined to move round the equator at a uniform rate, the same as that of the dynamical mean Sun, n degrees per day. The mean Sun is made to coincide with the dynamical Sun when it is at the First Point of Aries. In one year it will complete a circuit of the equator, and will once more be at the First Point of Aries. The position of this mean Sun, as given by its right ascension, can therefore be used to mark out time. In particular, the passage of the mean Sun across the South line on two successive occasions marks the passage of a mean solar day.

There will be a discrepancy between the time reckoned by the right ascension of the mean Sun and the time reckoned by the right ascension of the true Sun. The discrepancy is not constant, because of the apparent non-uniform motion of the true Sun. This discrepancy is recorded as "the equation of time" which is defined as

$$\text{Equation of time} = \begin{pmatrix} \text{Right ascension of} \\ \text{the mean Sun} \end{pmatrix} - \begin{pmatrix} \text{Right ascension of} \\ \text{the true Sun} \end{pmatrix}$$

5.6   RIGHT ASCENSION OF THE MEAN SUN

Let the time when the true Sun is at perigee $P = t_0$. Then the dynamical mean Sun is also at perigee at time t_0. The dynamical mean Sun moves along the *ecliptic* at the rate of $n = 360/365\frac{1}{4}$ degrees per day. It therefore moves from P to γ, the vernal equinox, in a time $(360 - \tilde{\omega})/n$ days, where $\tilde{\omega}$ is the longitude of perigee of the true Sun's orbit referred to the Earth as central body (see Figure 5.4). The dynamical mean Sun is therefore at γ at time t_1, where

$$t_1 = t_0 + \left(\frac{360 - \tilde{\omega}}{n}\right)$$

The mean Sun is therefore at γ at time t_1.

If we now consider the position of the mean Sun at any time t, that is, after an interval of $(t - t_1)$ from when it was at γ, we see that it has moved along the *equator* to M, such that $\angle \gamma OM = n(t - t_1)$. The angle γOM is the right ascension of the mean Sun.

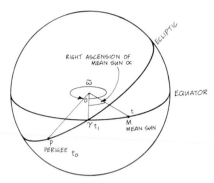

Figure 5.4   Right ascension of mean Sun

Thus,   right ascension of the mean Sun $= n(t - t_1) = n\left(t - t_0 - \dfrac{\overline{360 - \tilde{\omega}}}{n}\right)$

or,   right ascension of the mean Sun $a = nt - (nt_0 + 360 - \tilde{\omega})$.

5.7   PROJECT 24

To calculate the equation of time for various dates throughout the year, and to plot the equation of time throughout a period of one year.

Data

The Table below gives the right ascension of the true Sun for dates throughout 1969.

For each of the dates given above, complete Table 5.4 with columns as given below.

To assist in the completion of the Table, the following notes will be relevant:

(a)   Apart from the interval between the first two dates, the interval between successive dates is always 10 days. Column 2 will therefore proceed 0, 6, 16, 26, 36 etc.

(b)   The value of n, the mean daily motion, is

$$n = \frac{360}{365\frac{1}{4}} \text{ degrees per day}$$

(c)   Column 4 will be the constant value $(360 - 282.5)$ or 77.5 degrees.

TABLE 5.3

Date		Right ascension $a \odot$	
$t_0 =$ Jan	3	18h 54.1m	Longitude of perigee of
	9	19 20.4	the Sun's apparent orbit
	19	20 03.5	$\bar{\omega} = 282.5°$
	29	20 45.4	
Feb	8	21 25.9	Date of perigee
	18	22 05.2	$t_0 = 1969$ January 3
	28	22 43.4	
Mar	10	23 20.4	
	20	23 57.1	
	30	0 33.5	
Apr	9	1 10.0	
	19	1 46.9	
	29	2 24.5	
May	9	3 02.9	
	19	3 42.3	
	29	4 22.6	
Jun	8	5 0.36	
	18	5 45.1	
	28	6 26.7	
Jul	8	7 07.9	
	18	7 48.6	
	28	8 28.3	
Aug	7	9 07.1	
	17	9 44.9	
	27	10 21.8	
Sep	6	10 58.2	
	16	11 34.1	
	26	12 10.0	
Oct	6	12 46.2	
	16	13 23.1	
	26	14 00.8	
Nov	5	14 39.8	
	15	15 20.2	
	25	16 01.9	
Dec	5	16 44.9	
	15	17 28.8	
	25	18 13.2	
	27	18 22.2	

TABLE 5.4

(1) Date	(2) $t-t_0$ days	(3) $n(t-t_0)°$	(4) $(360-\bar{\omega})°$	(5) $a°$
Jan 3	0	0	77.5	−77.5
9	6	5.9	77.5	−71.6
19	16	15.8	77.5	−61.7

	(6) $a°$	(7) $a_\odot$	(8) $(a_m - a_\odot)$
Jan 3	−5h 10.0m	18h 54.1m	−4.1m
9	−4 46.4	19 20.4	−6.8
19	−4 06.8	20 03.5	−10.3

(d) Column 5 will be column 3 minus column 4 since right ascension mean Sun = $n(t - t_0) - (360 - \bar{\omega})$ degrees as shown previously.

(e) Column 6 is the right ascension of the mean Sun converted from degrees to hours and minutes. For this purpose we must remember that 360 degrees corresponds to 24 hours of time. Broken down further, this means that 15 degrees=1 hour, and 1 degree=4 minutes.

It is very important in this conversion to be as accurate as possible since column 8 is the difference of column 6 and column 7. These two values are often close to one another, and any error in one of the columns is greatly magnified when the difference is taken. Where the numerical value in column 6 is less than that in column 7, an addition of 24 hours should be made to the value of column 6 before the difference is calculated.

(f) Column 8 is, of course, the Equation of Time since
    Equation of time = R.A. mean Sun − R.A. true Sun.

Finally, plot on a graph date as abscissa and equation of time as ordinate to scales of   10 mm = 20 days   and   10 mm = 2 minutes of equation of time.

The time axis should be positioned to allow for both positive and negative values of the equation of time. Note that the graph crosses the horizontal axis at four dates, which means that at these dates the equation of time is zero. Note these dates, and compare them with April 16, June 4, September 1 and December 25. These dates vary only slightly from year to year.

5.8   THE ANALEMMA

The word 'analemma' has similar derivations from both the Latin and the Greek languages. In the former it meant the pedestal on which a sundial was placed, and in the latter simply a prop. The word is scarcely used at all today, but it was used in the past to describe a curve in the form of a figure eight normally seen on globes of the Earth.

Not surprisingly, in its modern usage it is related to the sundial in that its co-ordinates parallel to the equator give the equation of time at any date throughout the year.

Co-ordinates in the north–south direction give the declination of the Sun at that date. It will be seen that if the top and bottom of the figure of eight are made to touch the circles on the globe representing the Tropic of Cancer and the Tropic of Capricorn respectively, then the declination of the Sun at any date can be seen from the corresponding latitudes on the globe, either north or south. Also, if the analemma were so placed on the Greenwich meridian (longitude 0°) it would show the geographical latitude of the sub-solar point at noon by Greenwich mean time for each day of the year. The sub-solar point is that point on the Earth's surface at which the line through the Sun normal to the Earth's surface cuts the surface of the Earth.

The existence of an equation of time is due both to the eccentricity of the Earth's orbit round the Sun and to the inclination of the Earth's axis to the plane of the Earth's orbit. Since both of these are slowly changing with time, the shape of the analemma is also changing very slowly.

5.9   PROJECT 25

To construct the shape of the analemma for a given year.

The Table below gives the declination of the Sun and the equation of time for various dates in the year 1969.

Set up axes parallel to the edges of the paper with the origin near the centre of the page. Mark the horizontal axis 0 to 20 minutes of time to a

TABLE 5.5

	Date	Equation of time	Declination of the Sun
January	10	−7.38m	−22.02°
	20	−10.95	−20.20
	30	−13.28	−17.77
February	10	−14.28	−14.48
	20	−13.87	−11.05
	28	−12.72	−8.12
March	10	−10.48	−4.27
	20	−7.70	−0.32
	30	−4.68	+3.62
April	10	−1.47	+7.80
	20	+0.98	+11.38
	30	+2.75	+14.65
May	10	+3.65	+17.52
	20	+3.60	+19.90
	30	+2.65	+21.72
June	10	+0.80	+22.98
	20	−1.32	+23.42
	30	−3.42	+23.20
July	10	−5.17	+22.20
	20	−6.23	+20.75
	30	−6.37	+18.62
August	10	−5.37	+15.68
	20	−3.47	+12.58
	30	−0.77	+9.15
September	10	+2.85	+5.10
	20	+6.22	+1.25
	30	+9.82	−2.65
October	10	+12.83	−6.50
	20	+15.06	−10.20
	30	+16.27	−13.65
November	10	+16.08	−17.03
	20	+14.45	−19.60
	30	+11.50	−21.58
December	10	+7.40	−22.88
	20	+2.62	−23.43
	25	+0.15	−23.42
	26	−0.35	−23.38

scale of 10 mm = 4 minutes to represent the equation of time on both sides of the origin. Mark the vertical axis 0 to 24 degrees to a scale of 10 mm = 2.5 degrees to represent the declination of the Sun, also on both sides of the origin.

From the Table read corresponding values of the equation of time and the declination of the Sun, and plot the points they represent on the paper.

Draw a smooth curve through these points taking them in chronological order. The curve will be seen to be in the form of a figure of eight, points on the right of the vertical axis being the Sun's declination when the sundial time is in advance of the local mean time.

For dates not included in the Table, the declination and equation of time can be obtained from an estimated point lying between the two nearest dates given in the Table.

The analemma is, of course, not very accurate due to the smallness of the scales. It is particularly inaccurate where the curve is vertical or horizontal. Nevertheless the analemma gives a pictorial view of the variation of the two quantities involved.

5.10 SYNCHRONOUS ORBITS

At this point we return briefly to the orbit of an artificial satellite around the Earth. Referring to the nomograph of Project 8 we see that an artificial satellite launched in a circular orbit at a distance 46 168 km

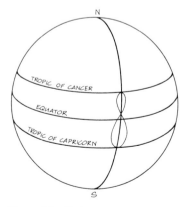

Figure 5.5   Project 25. Analemma for 1969

from the centre of the Earth will have a period of exactly 1 day. If the orbit is exactly above the equator and moving in the same direction as the Earth's equator, then the satellite will remain above the same point on the Earth's surface. Such synchronous satellites can therefore be used as communication satellites.

If however the radius of the orbit is the same but the inclination of the orbit to the equator is, say 30°, then the path of the satellite will be a figure of eight oscillating between latitudes ± 30°. The path traced out is very similar to the analemma just described and the reasons for the figure of eight are the same, except that in the one case the time scale is 1 year and in the other it is 1 day.

5.11   THE MERIDIAN

Referring to the celestial sphere again (Figure 5.6), PP' is the axis of the Earth's rotation, P being the north celestial pole. Z is the observer's zenith, that is, the point immediately above his head, and X is the Sun.

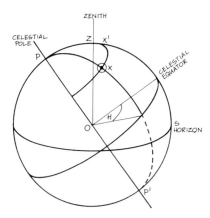

Figure 5.6   The meridian and hour angle

As the Earth rotates throughout the day, the Sun will appear to travel in the sky in a path which is parallel to the equator. When it arrives at X', it will be furthest from the horizon, and the plane through PZX' will cut the horizon in S, which we call south.

The plane PZX' cuts the celestial sphere in a circle called the meridian.

5.12   THE HOUR ANGLE OF THE SUN

Since the plane PXP' sweeps out equal angles in equal times due to the uniform rate of rotation of the Earth, the angle H may be taken as a measure of time. H is called the hour angle of the Sun.

The time thus marked out is referred to as *apparent* solar time. As we have seen, due to the non-uniform motion of the Sun in the ecliptic from day to day, this method of measuring time is not used today, and the mean solar time described in Section 5.5 has replaced it.

5.13   THE SUNDIAL

Although apparent solar time is not used generally today, it is nevertheless the time shown by sundials, many of which can still be seen on churches and in gardens. When used with the equation of time a sundial can give mean solar time, the time used in everday life.

Any sundial consists essentially of a plate on which are marked the hours, and a style which casts a shadow on the plate. The plate may be flat, usually horizontal or vertical, or cylindrical, or even a circular narrow band. The style is often triangular but can be simply a rod. The edge of the style which forms the shadow on the hour divisions always points towards the north celestial pole (Figure 5.7).

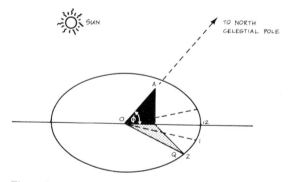

Figure 5.7   Sundial

5.14   PROJECT 26

To make an equatorial sundial.
The simplest form of equatorial sundial consists of a style parallel to the axis of the Earth, that is pointing towards the Pole Star, and a flat

circular base at right angles to it, that is parallel to the Earth's equator. Since, due to the rotation of the Earth, the Sun moves 15° per hour around the style, the hourly markings on the base are equally spaced at 15° intervals (see Figure 5.9(c)). This equatorial sundial will be used later in Project 27 to develop the markings on a sundial with a horizontal base.

A more attractive equatorial sundial may be made if, instead of a circular flat plate we imagine a hollow circular cylinder to be erected on it so that the style is the axis of the cylinder. The hour markings will then be those generators of the cylinders starting at the points where the hour markings on the base plate cut the cylinder. To allow the shadow of the style to be thrown on to the inside of the cylinder, half of the cylinder on the south side is removed.

Figure 5.8 shows a suitable layout, the height of the cylinder being about 150 mm. The cylinder itself may be made of stiff but flexible card on which are drawn the hour lines. The ends of the cylinder may be wooden discs and a stiff metal or plastic rod such as a knitting needle can form the style. The base can be a wooden block cut at the appropriate angle shown in Figure 5.8.

5.15   PROJECT 27

To make a model to illustrate how the markings on a horizontal sundial may easily be obtained.

The markings on a horizontal sundial are not spaced at equal intervals and thus the construction of them for any given latitude might seem to present some difficulty.

As Project 26 has demonstrated, the equatorial sundial, however, presents no such difficulty since it has its plane at right angles to the polar axis and its style parallel to the polar axis. The hour markings are uniformly spaced at 15° intervals since the Earth rotates uniformly about this axis through 360° in 24 hours.

The model shown in Figure 5.9(a) indicates how the transfer of markings from the equatorial dial to the horizontal dial can be effected. It consists of two boards or cards A and B hinged along the line YY'. C is the style for both dials, and is arranged so that it is at right angles to A, the equatorial dial, and makes an angle φ with the horizontal dial, where φ is

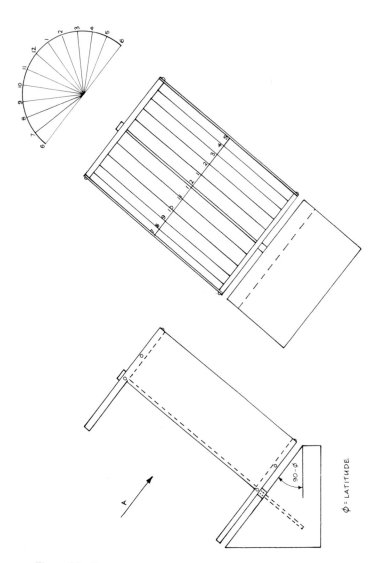

Figure 5.8   Equatorial sundial

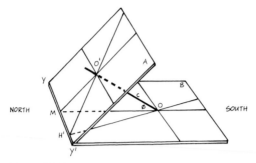

Figure 5.9(a)

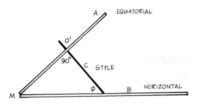

Figure 5.9(b)

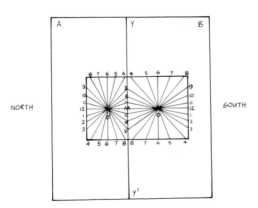

Figure 5.9(c)

Figure 5.9   Sundial demonstration model

the latitude of the place where the dial is to be used (Figure 5.9(b)). It is only necessary to determine two points on each hour marking on either board to determine that marking, which will be a straight line. On the equatorial O' is one point and on the horizontal O is one point on all hour lines. If $O'H'$ is an hour line on the equatorial, then H' also lies on the corresponding hour line on the horizontal dial. Thus each hour line is identified on the horizontal sundial.

With this principle established we can imagine the lines on A to be on the reverse side, and then fold A down towards the north so that it is in the same plane as B (Figure 5.9(c)). The construction of the horizontal dial is now obvious.

5.16   PROJECT 28

To make a simple pocket sundial.

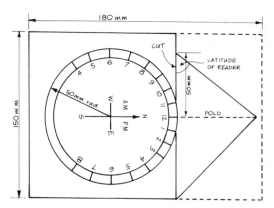

Figure   5.10   Plan of pocket sundial

Cut a thin card   180 mm × 150 mm.   Mark it out as in Figure 5.10. The angles between the hour graduations may be obtained from Project 27. Alternatively, they may be taken from the Table below:

Cut and fold the card as in Figure 5.11.

To operate as a sundial the base plate must be horizontal, and the radius 0 – 12 directed to the north.

TABLE 5.6

	Angles
Noon and 1 p.m. and also noon and 11 a.m.	12°
Noon and 2 p.m. and also noon and 10 a.m.	24.5
Noon and 3 p.m. and also noon and 9 a.m.	38
Noon and 4 p.m. and also noon and 8 a.m.	54
Noon and 5 p.m. and also noon and 7 a.m.	71
Noon and 6 p.m. and also noon and 6 a.m.	90
Noon and 7 p.m. and also noon and 5 a.m.	109
Noon and 8 p.m. and also noon and 4 a.m.	126
Noon and 9 p.m. and also noon and 3 a.m.	142
Noon and 10 p.m. and also noon and 2 a.m.	136
Noon and 11 p.m. and also noon and 1 a.m.	169

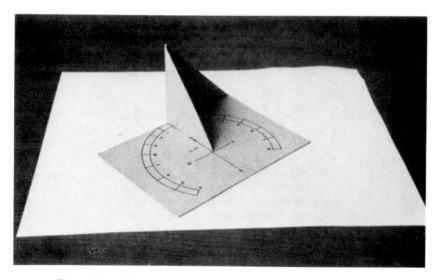

Figure 5.11   Folded pocket sundial

5.17   SUNSPOTS

Occasionally, when the Sun is shining through a fog or mist its brilliance is reduced so that the disc looks a dull red, and a small dark patch

may be seen on the Sun's surface. This is a sunspot. It is not a perma-
nent feature of the Sun's disc, but may last from a few days to a few
weeks. The surface of the Sun is very hot, and sunspots are areas of
relatively cool gas connected with magnetic fields on the Sun. Compared
with the rest of the Sun they therefore look dark.

Sunspots are often present on the Sun, but remain undetected by the
casual observer because of the brilliance of the Sun. By projecting the
image of the Sun on to a screen by means of a telescope, the light
intensity is reduced, and sunspots can be seen more comfortably. Readers should note that it is extremely dangerous to look at the Sun directly
through a telescope, and this should *never* be done. Blindness can result
in a matter of seconds.

If a count is made of all the sunspots observed over a period of time, it
is found that the number appearing fluctuates. Sunspot activity seems to
reach a maximum about every 11 years.

There is a tendency for sunspots to form near the belts corresponding to
latitudes of about 40° north and 40° south at the beginning of the sunspot cycle. The formation of spots moves nearer to the Sun's equator as
the cycle progresses, being mainly in latitudes 15° north and 15° south at
maximum sunspot activity, and then continues to move nearer the
equator to the end of the cycle.

Individual sunspots give evidence that the Sun is rotating. They move
across the disc, and sometimes reappear as one revolution of the Sun is
completed. Also, those sunspots nearer the Sun's equator move more
rapidly than those further away. In fact, one revolution at the equator
takes about 27 days, while for belts nearer the poles the time of revolution can be 3 or 4 days more than this.

5.18   PROJECT 29

To plot the sunspot number against date, and to deduce the period of
the sunspot cycle.

The number of sunspots observed will depend upon the telescope being
used. A number, known as the Wolf number, is associated with each
photograph of the Sun, and is an indication of the number of sunspots
on the photograph. Each sunspot observed in a particular telescope as
standard is counted as unity, and each group of sunspots as ten. The
total is then multiplied by a factor to take account of the telescope being
used.

Data

The Table below gives the sunspot numbers for each year over a period of 50 years.

TABLE 5.7

Year	Number	Year	Number
1910	31	1935	59
1911	15	1936	100
1912	11	1937	125
1913	7	1938	128
1914	22	1939	113
1915	49	1940	82
1916	74	1941	62
1917	99	1942	42
1918	90	1943	27
1919	69	1944	20
1920	41	1945	50
1921	33	1946	100
1922	20	1947	156
1923	12	1948	144
1924	23	1949	150
1925	69	1950	91
1926	87	1951	66
1927	110	1952	53
1928	79	1953	25
1929	74	1954	16
1930	52	1955	53
1931	29	1956	163
1932	15	1957	220
1933	11	1958	225
1934	19	1959	204

Plot the sunspot number as ordinate, using a scale 10 mm = 20 units, and the year as abscissa, using a scale 20 mm = 5 years. Observe the peaks and the distances in years between them. Take an average of the peak to peak times.

Compare your value with the value of 11.1 years deduced by Wolf. It will be noticed that the recurrence of the sunspot cycle has been nearer 10 years over the period plotted, namely 50 recent years.

Looking at the sunspot diagram it would also appear that there is a secondary cycle of a much greater period superimposed, but the data given has not been for a sufficiently long period to confirm this.

5.19   PROJECT 30

(i)   To construct Stonyhurst Discs to enable the latitude of sunspots to be determined.

(ii)   To determine the latitude of a number of sunspots.

The axis of rotation of the Sun is not perpendicular to the plane of the Earth's orbit, but is inclined to it at about 7.2°. Thus, when the Earth is on one side of the Sun, the axis of the latter may be tilted directly away from the Earth at, say, the north end. When the Earth is in the diametrically opposite position, the north end of the Sun's axis would then be tilted directly towards the Earth at the same angle of 7.2°. For other positions of the Earth, the angle of tilt towards or away from the Earth will be less than this. This angle is usually given the symbol B_0, and its value for any date in any year is listed in publications such as the *Handbook of the British Astronomical Association*.

In a similar way, the axis of the Sun will appear to swing slowly from side to side about the north point as the Earth travels in its orbit. Here the tilt of the Earth affects the apparent angle, and the swing on each side of the north point is about 26.4°. This angle is usually designated P, and is also listed in the above publication for various dates throughout the year.

When the axis of the Sun is tilted directly towards or directly away from the Earth, the lines of latitude will have maximum distortion, being curved downwards in the former case and upwards in the latter case (Figure 5.12). There will be only two dates in each year when the lines of latitude appear as straight lines, namely when the plane containing the Sun's axis and the north–south axis is at right angles to the line joining the Sun to the Earth (Figure 5.13).

Thus the determination of the latitude of sunspots is not as simple as might be imagined at first sight.

To enable the latitude of sunspots to be read off directly from photographs, a useful device, known as the Stonyhurst Disc, has been devised. This is a disc with lines of latitude drawn for a particular angle B_0. It is not practical to have a set of Stonyhurst Discs for all possible values of

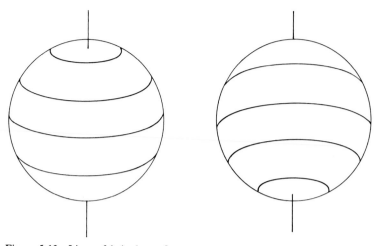

Figure 5.12   Lines of latitude on Sun.
(a) Sun's axis titled away from Earth (b) Sun's axis tilted towards Earth.

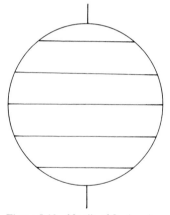

Figure 5.13   No tilt of Sun's axis

B_0, and it is usual to compromise by having a set of discs for, say, $B_0 = 0$, $2°$, $4°$, $6°$, and $7°$, intermediate values being estimated.

The present project takes a photograph of the Sun, and constructs a single Stonyhurst Disc corresponding to the value of B_0 relevant to the photograph. It is simply an exercise in projection from the disc for $B_0 = 0$, which is easy to draw.

Take two centres A and B 125 mm apart. On each centre draw a circle of radius equal to that of the Sun in the photograph (Figure 5.14). Draw the diameter *DE* of the circle with centre *A* which makes 6.6° upwards with the line *AB*. This angle is the value of B_0 for the date of the photograph. Draw the diameter *FG* at right angles to the first. The two diameters represent respectively the equator and polar axis for the disc $B_0 = 0$.

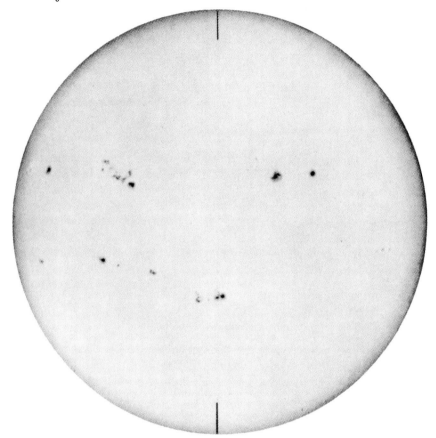

Figure 5.14   Sunspots (by kind permission of Philip J. del Vecchio, New Jersey, U.S.A.)

From *A* draw radii at 10° intervals for the right hand half of the circle centre *A*. A typical radius will cut the circle at *C*. From *C* draw parallel to the equator. This represents the line of latitude for the angle *CAE*. From *C* draw parallel to *FG* to cut *DE* in *H*. Then *H* is the point on *DE* corresponding to longitude $(90° - C\hat{A}E)$. Repeat for all the lines of latitude, and all the longitude points on *DE* spaced at 10° intervals.

With centre A and radius AH, draw a semi-circle above DE. For all points where the radii cut this circle, drop perpendiculars on to the latitude through C. These points will divide the latitude line through C into 10° longitude intervals. Repeat for all such associated pairs of points H and C. Join up corresponding sets of points to form the lines of longitude. Label these 9, 8, 7 . . . 0 successively from A.

Centrally below the circle centre B, draw 7 parallel lines horizontally spaced about 5 mm apart. Let the centres of these lines be labelled O. On the first line (0°) mark off distances 01, 12, 23 . . . 89 corresponding to 98, 87, 76 . . . 10 on the 0° latitude line on the circle centre A. Mark off similarly on the other lines corresponding distances on the 10°, 20° . . . 60° latitudes.

Taking a typical point a on the grid of latitude and longitude on circle centre A (in this case, latitude 20° north, longitude 50°, reference line 4) draw horizontally from a and vertically from 4 on the horizontal line 20°, which was drawn below circle centre B, to intersect in a'. Then a' lies on latitude 20° on the Stonyhurst Disc for $B_0 = 6.6°$. Repeat for all other points on the 20° latitude line on circle centre A, and so obtain a set of points which, when joined up, will be the line of 20° latitude north on circle centre B. Repeat for other latitudes from 0° to 60° both north and south.

This is sufficient for our purpose, and if diagram $B_0 = 6.6°$ is overlaid with the tracing of the photograph of the Sun with the sunspots marked on it, it will be a simple matter to read off the latitude of each spot.

For completeness of the grid, the lines of longitude can be copied from the circle centre A to the circle centre B, although it should be realised that, as the Sun is rotating, longitude from these diagrams has little meaning.

5.20   REFRACTION DUE TO THE ATMOSPHERE

The light coming from the Sun to an observer on the surface of the Earth passes through the atmosphere, which bends the light in such a way that the observer appears to see the Sun nearer the zenith than it really is. Since the upper layers of the atmosphere are less dense than those below the bending of the light rays is progressive. The effect is small when the Sun is high in the sky but appreciable when the Sun is near the horizon since the passage of the rays is then mostly through the denser part of the atmosphere.

The same is true for stars and an allowance is made for refraction when the true position of a star is being determined.

The next Project 31 assumes that the atmosphere is divided into five layers only, whereas the variation in atmosphere density is continuous. The refractive indices have also been exaggerated but the general result deduced is true for the real atmosphere.

Demonstration 2 shows that due to the effect of refraction the Sun can apparently be seen above the horizon when it is actually below it.

5.21 PROJECT 31

To plot the path of a ray of light as it passes through layers of different and increasing refractive indices and to determine the apparent displacement of a star observed through these layers.

Referring to Figure 5.15, the relation between the angle of incidence i_0 and the angle of refraction r_1 of the ray of light at the junction of two layers is

$$\frac{\sin i_0}{\sin r_1} = \frac{\mu_1}{\mu_0}$$

or

$$\sin r_1 = \frac{\mu_0}{\mu_1} \sin i_0$$

where μ_0, μ_1 are the refractive indices for layers 0 and 1 respectively.

Draw five layers each 50 mm thick, with refractive indices 1.0, 1.1, 1.3, 1.6 and 2.0 respectively from the top. Let the junction between the layers be AB, CD, EF, GH respectively. Let M be the mid-point of AB. Draw the normal NMN_1 at M.

We shall take i_0 to be 30° in this case for the sake of example.

Then

$$\frac{\sin 30°}{\sin r_1} = \frac{1.1}{1.0}$$

or

$$\sin r_1 = \frac{0.5}{1.1} = \frac{0.455}{1}$$

With centre M and radius 25 mm draw an arc. By trial place a scale at right angles to MN_1 such that the perpendicular distance from the arc to MN_1 is $0.455 \times 25 = 11.4$ mm. Mark this point on the arc, say P. Join M to P and produce it to cut CD in M_1.

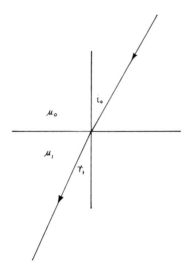

Figure 5.15   Refraction at a surface

For the junction CD, r_1 becomes i_1, so that

$$\frac{\sin r_1}{\sin r_2} = \frac{1.3}{1.1}$$

or

$$\sin r_2 = \frac{1.1}{1.3} \sin r_1 = 0.455 \times \frac{1.1}{1.3} = \frac{0.385}{1}$$

As before, draw an arc of 25 mm radius from M_1 and adjust the scale so that the perpendicular distance from the arc to the normal through M_1 is $0.385 \times 25 = 9.6$ m. Continue this process down to the bottom layer, the position of the observer, M_3.

Measure the angle a between the normal at M_3 and $M_3 P_3$. $M_3 P_3$ is the direction in which the light from the star appears to come to the observer.

Calculate $\sin 30°/\sin a$ and compare this with the refractive index of the lowest layer, showing that the apparent displacement of a star towards the zenith may be calculated if the refractive index of the air at the observer only is known.

In practice this is about 1.0003, the values above having been exaggerated.

5.22 DEMONSTRATION 2

Demonstration on the setting Sun

For this demonstration we shall need a book or books of total thickness about 50 mm, a clear glass wine bottle about 70 mm diameter and with parallel sides. The bottle should be completely filled with clean water to represent the Earth's atmosphere, and corked. In addition we shall need to make up a thin card about   80 mm × 60 mm,   bent parallel to the longer edges so that the card can be free-standing, presenting a vertical area   80 mm × 45 mm   as in Figure 5.16. In the centre of the vertical area a circular hole 10 mm diameter should be cut to represent the Sun. For better effect cover the back of this hole with thin red or orange tissue paper.

Arrange these items as shown in Figure 5.17. Look along the surface of the book, which represents the horizon, and the image of the Sun will be clearly seen, even though the whole of it is below the horizon. Remove the bottle and the image of the Sun cannot be seen.

This demonstration is effective in daylight.

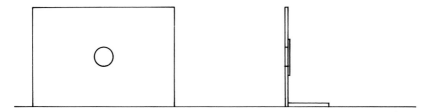

Figure 5.16   Artificial sun

5.23   THE COLOUR OF THE SKY

The Earth's atmosphere contains a large number of fine particles. The light from the Sun is made up of the colours of the rainbow and in passing through the clouds of fine particles the blue colours are scattered more than the red colours. Thus the Sun looks red when it is low in the sky, near the horizon. The blue light is scattered and the red light penetrates the denser layers of the atmosphere.

The foregoing mechanism is also responsible for the blue appearance of the sky in the daytime since the light from the Sun has its blue components spread in its passage through the atmosphere.

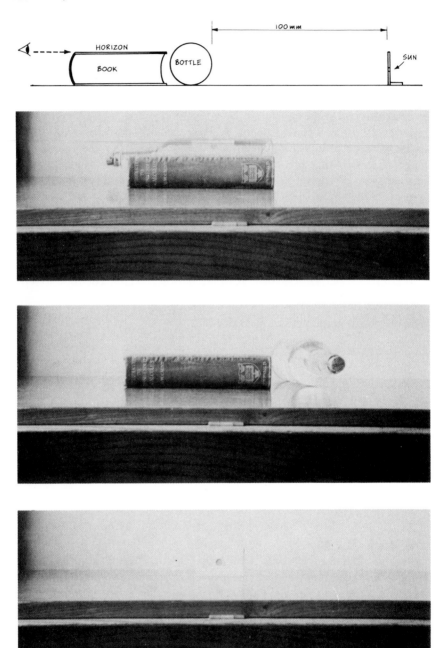

Figure 5.17   Refraction demonstration

5.24 DEMONSTRATION 3

Sunset and blue sky demonstration

For this demonstration we shall need a rectangular glass tank of approximately 5000 ml capacity. A standard glass aquarium tank can be used. Alternatively, since acid is going to be added, a tank can be made up from sheet glass stuck together and sealed with silicone rubber, a tube of which can be bought at almost any shop dealing in tropical fish. A tank 150 mm × 150 mm × 250 mm has been found convenient and satisfactory.

The tank is filled with a solution of hypo (sodium thiosulphate) to a strength of about 100 g of hypo to 5000 ml of water. A 35 mm slide projector, in which the special slide *A* (see Figure 1.3) described in Section 1.3 has been placed, and a white screen are arranged so that the light is projected through the hypo parallel to the longest sides of the tank to form a bright disc on the screen to represent the Sun (Figure 5.18). The room is darkened and a few drops of concentrated hydrochloric acid are added to the hypo. A total of about 0.5 ml of acid will be found suitable.

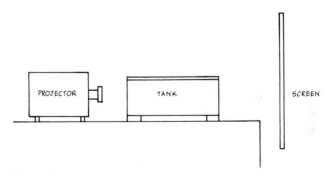

Figure 5.18   Sunset demonstration

A colloidal suspension of sulphur slowly forms, rising from the bottom of the tank. As the suspended particles increase in size, and the light components are scattered, the disc of light on the screen slowly becomes yellow, orange and finally red giving a very realistic display similar to that of the setting Sun.

If the scattered light coming from long sides of the tank is observed, it will be seen to be blue. It is also polarised, a term which will not be defined here, but which can be demonstrated by looking at the side of the tank through a piece of polaroid. If the polaroid is rotated slowly the light will be seen to go dark and light alternately.

The rate of progress of the artificial sunset can be speeded up or slowed down according to whether more or less than 0.5 ml of hydrochloric acid are added.

The chemical reaction is represented by the equation

$$Na_2S_2O_3 + 2HCl \rightarrow 2NaCl + SO_2 \uparrow + H_2O + S \downarrow$$

A much safer, but less effective, alternative is to use water in the tank and add a very small quantity of instant non-fat dried milk.

5.25   NUCLEAR REACTIONS IN THE SUN

The Sun is a star which is giving out energy in the form of radiation at a rate of about 4×10^{26} J/s. According to geological evidence which concerns radioactive elements, this rate of radiation has been maintained for at least 10^9 years. If the energy of the Sun were being produced by the normal burning of some fuel such as the heating oil which we use on Earth, the energy it would produce per kilogram would be about 45 MJ. The mass of the Sun is about 2×10^{30} kg. A simple calculation will show that under these conditions the whole of the Sun would be consumed in about 7000 years. Clearly the energy must come from some other processes, which are nuclear reactions taking place in the central part of the Sun.

We are accustomed to atomic and hydrogen bombs and the speed and violence of the nuclear processes in them. The nuclear reactions in the Sun, however, take place slowly and, on average, a nucleus at the centre of the Sun is involved in a reaction once in about 3×10^9 years.

With each reaction, energy is emitted or absorbed in the form of photons travelling at the speed of light. Any one photon, therefore, will follow a tortuous path within the Sun before it can be radiated at the Sun's surface, if indeed it can be identified as the same photon! As each photon is absorbed, another photon may be re-emitted in quite a different direction and, in turn, this will travel, again on average, about 10 mm before it is itself absorbed and re-emitted in yet another direction.

We might think that since photons are re-emitted in random directions they will eventually end up back at the centre of the Sun. This is true on average. However, Appendix A8 shows that there is a probable number of steps for a photon to achieve a definite distance from the Sun's centre, and this distance can be the radius of the Sun. This photon will then have escaped from the Sun and become radiation from the Sun's surface. Appendix A8 shows that the time taken for this to occur is about 6000 years.

A convincing demonstration, best carried out with pairs of students, and a variation of the basic demonstration which can be used in front of a large audience will now be described.

5.26 DEMONSTRATION 4

Energy transport in the Sun demonstration

(a)   For this two-dimensional version of the demonstration, each pair of students will require a dice, a container for shaking it, a pencil and a sheet of isometric graph paper, or a sheet ruled in a similar manner (see Figure 5.19).

Using the mid-point O of the isometric graph paper as centre, draw a circle whose radius is equal to half the width of the graph paper. This circle then represents the surface of the Sun, and each point of intersection of the lines within the boundary represents a possible position of emission of a photon. It will be seen that each point of intersection lies at the centre of a hexagon, any one of the vertices of which could be the next point of involvement of the photon which is at present at its centre. The direction of these vertices from the centre of the hexagon can be represented by the numbers 1 to 6 on the dice. The distance between any two adjacent points can be taken as the mean free path of a photon.

The demonstration will proceed more smoothly if one student uses the pencil and the other student throws the dice.

Starting with the pencil at O, throw the dice and draw a line from O to the next vertex corresponding to the number thrown. The new pencil position then represents the centre of a similar hexagon, and the process is repeated until the pencil crosses the boundary of the Sun, the line drawn being the random path of the photon from centre to boundary of the Sun. The time taken or the number of throws required should be recorded. Either of these will be an indication only of the unexpectedly large time it takes for a photon of energy to travel from the centre to the boundary of the Sun since this demonstration is for a very limited

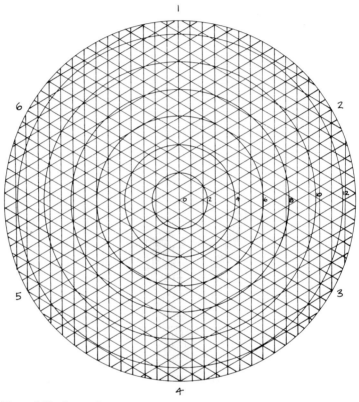

Figure 5.19   Isometric grid

radius, the motion is two-dimensional and the emissions are restricted to one of only six directions. Furthermore, it is assumed that the time between the absorption and re-emission of a photon is small compared with the travelling time of a similar photon.

(b)   For demonstrating the random passage of a photon from the centre to the surface of the Sun in such a way that all members of an audience can see, an overhead projector can be used. We shall need an isometric grid of holes, corresponding to the intersections of the lines on the isometric graph paper in (a), but drilled in thin sheet aluminium (Figure

5.20). The holes will terminate on the circular boundary of the Sun. It has been found that holes of about 1.5 mm diameter are satisfactory, and that the distance between hole centres, measured in the directions of the base hexagon, can profitably be 6 mm. The position of the photon at any time can be seen on the screen if the corresponding hole is covered by a small circular disc of coloured, translucent perspex. The diameter of the

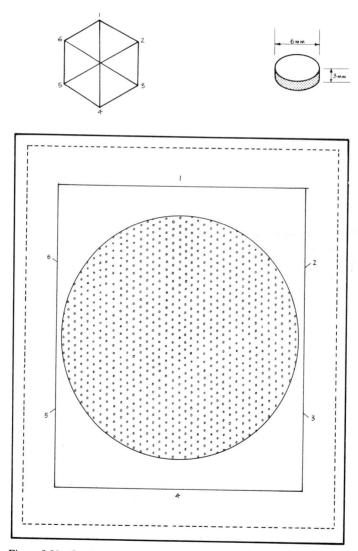

Figure 5.20   Overhead projector grid

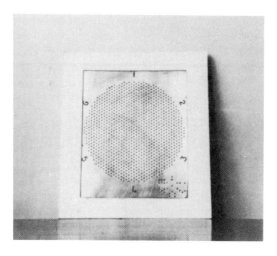

Figure 5.20   continued – overhead projector grid

disc can be about 6 mm so that only one hole is covered at a time. The thickness of the disc is not important provided that light passes through easily, and 3 mm has been found to be acceptable.

The aluminium sheet itself can be 300 mm × 270 mm, and should preferably be put in a standard cardboard frame normally used for overhead projector transparencies. This will allow a Sun's radius of about 100 mm. The numbers 1 to 6 can be marked in appropriate positions on the plate or cardboard frame. The disc can be moved by the tip of a pencil so that the hand does not obscure the field in the region of the disc.

The method is then exactly the same as in (a) using the throw of a dice for selecting the direction of movement of the disc.

(c) If the demonstration is carried out a number of times we can try to relate the number of throws required to achieve a given distance from the centre.

Modify the isometric drawing or the aluminium sheet by scribing circles, each with centre O of radius 2, 4, 6, 8, 10 and 12 units respectively.

During each trial note the number of throws of the dice to achieve each of these radii. Take an average for the trials for each radius and compare with 2^2, 4^2, 6^2, 8^2, 10^2 and 12^2 (see Appendix A8).

5.27   THE RISING AND THE SETTING OF THE SUN

We are taught very early that the Sun rises in the east and sets in the west. This is only *generally* true. The Sun rises *exactly* in the east and sets *exactly* in the west on two dates only, namely the vernal, or spring equinox (March 21) and the autumnal equinox (September 21).

On dates between March 21 and September 21 the Sun rises north of east and sets north of west. Between September 21 and March 21 following the Sun rises south of east and sets south of west.

The Sun rises the furthest north of east and sets the furthest north of west when its declination is largest at +23°26′ approximately. The date then is that of the summer solstice, June 21. Similarly, the Sun rises the furthest south of east and sets the furthest south of west when the declination of the Sun is −23°26′ approximately. The date then is December 21, the winter solstice. On other dates the Sun rises and sets between these limits. The latitude of the observer will affect how far from the east the Sun rises and how far from the west the Sun sets.

5.28   PROJECT 32

To construct a nomograph relating the latitude of the observer φ, the declination of a heavenly body δ and the angle a from the east at which a body will rise or from the west will set.

The effect of refraction due to the atmosphere is to make a heavenly body appear nearer the zenith than it really is. This effect is ignored in the following.

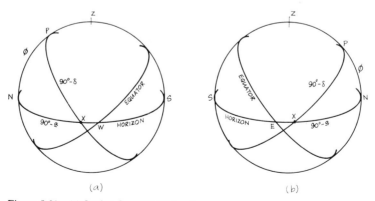

Figure 5.21   (a) Setting Sun. (b) Rising Sun

It is shown in Appendix A4 that for a body on the horizon, that is either rising or setting, φ, δ and a are related by

$$\sin a = \sec \varphi . \sin \delta$$

Draw the horizontal x axis across the middle of a sheet of graph paper. Set up two vertical axes 100 mm apart, stretching 100 mm above and 100 mm below the x axis. The origin is to be taken at the intersection of the left hand vertical axis and the horizontal x axis.

Plot the graphs

  (i) $x = 1/(1 + \sin a)$, $y = 0$ for $0 \leqslant a \leqslant 90°$

 (ii) $x = 0$, $y = \cos \varphi$ for $0 \leqslant \varphi \leqslant 90°$

(iii) $x = 1$, $y = -\sin \delta$ for $-90° \leqslant \delta \leqslant 90°$

Use for all scales   10 mm = 0.1 unit of each variable.

A straight line drawn from the observer's latitude to the declination of the heavenly body will cut the horizontal axis in an angle a, which is the angle north of east at which the body rises or the angle north of west at which the body sets. Alternatively, if the body rises south of east and sets south of west, the nomograph operated as above will give these angles.

5.29   THE RISING AND THE SETTING OF THE MOON

The relation between a, δ and φ given in Appendix A4 still applies for the Moon on the horizon. For a given latitude φ, it will be seen from $\sin a = \sec \varphi . \sin \delta$ that a will be greatest when δ has its greatest possible value.

The motion of the Moon is more complex than the motion of the Sun. Whereas the Sun repeats its cycle of rising and setting in one year, the Moon takes 18.61 years to repeat its cycle completely. Since the phases of the Moon tend to complicate consideration of the Moon's cycle, it is best to consider how the risings and settings of the *full* Moon vary. Figure 5.22, after C. A. Newham, *The Astronomical Significance of Stonehenge*, shows the horizon with four sectors, 1 to 4.

When the full Moon rises at its extreme position P_1 in Sector 1 it will set at P_4 in Sector 4. The next full Moon will rise a little nearer to the east and set a little nearer to the west, say at Q in Sector 1 and Sector 4 respectively. Successive full Moons will continue this trend until, about six months later, it is rising at P_2 in Sector 2 and setting at P_3 in Sector 3.

The next six months will see it rising successively back to Sector 1, but it will not quite achieve the position P_1 but some point T. During the next year the process will be repeated, starting at T, moving to U and then back to V in Sector 1. After about 9 years the rising will be at W_1 in Sector 1 and the setting will be at W_4 in Sector 4. This is also an extreme position. Thus the next nine years or so see the Moon's rising and setting working its way back gradually from W to P. The whole of this cycle takes 18.61 years.

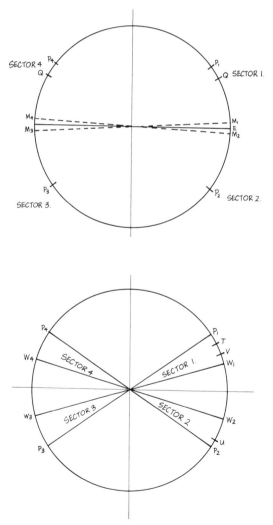

Figure 5.22 Positions of Moon rising and setting

5.30   STONEHENGE ALIGNMENTS

It is thought that the stones and posts which made up the complete Stonehenge were deliberately placed so that by standing in certain well-defined positions and looking towards certain other well-defined positions one would see certain astronomical events.

These events are given below.

A. (i)   Sunrise when furthest north of east – midsummer sunrise, 21 June

   (ii)   Sunset when furthest north of west – midsummer sunset, 21 June

  (iii)   Sunrise when due east – autumnal equinox, 21 September and vernal equinox, 21 March

  (iv)   Sunset when due west – autumnal equinox, 21 September and vernal equinox, 21 March

   (v)   Sunrise when furthest south of east – midwinter sunrise, 21 December

  (vi)   Sunset when furthest south of west – midwinter sunset, 21 December

B.   (i)   Full Moon rise when furthest north – point P_1, Figure 5.22 – midwinter moonrise

   (ii)   Full Moon set when furthest north – point P_4 – midwinter moonset

  (iii)   Full Moon rise when furthest north – point W_1 – midwinter moonrise

  (iv)   Full Moon set when furthest north – point W_4 – midwinter moonset

   (v)   Full Moon rise at its equinox – point M_1 – northern limit

  (vi)   Full Moon set at its equinox – point M_4 – northern limit

 (vii)   Full Moon rise at its equinox – point M_2 – southern limit

(viii)   Full Moon set at its equinox – point M_3 – southern limit

  (ix)   Full Moon rise when furthest south – point P_2 – midsummer moonrise

   (x)   Full Moon set when furthest south – point P_3 – midsummer moonset

  (xi)   Full Moon rise when furthest south – point W_2 – midsummer moonrise

 (xii)   Full Moon set when furthest south – point W_3 – midsummer moonset

5.31   PROJECT 33

To check the alignments given in A and B previously

Referring to the scale drawing of Stonehenge, Figure 5.23, as it was thought to be when completed, the alignments given below have been discovered.

A.   (i)   Position 91 seen from 92
            Heelstone seen from the centre
            Position 94 seen from 93
            Heelstone seen from the gap between Sarcen Stones 30 and 1

     (ii) Position 94 seen from hole *G*
            The gap between Sarcen Stones 23 and 24 seen from the gap between Trilithons 59 and 60

     (iii) Hole *F* seen from 93
            Hole *C* seen from 94

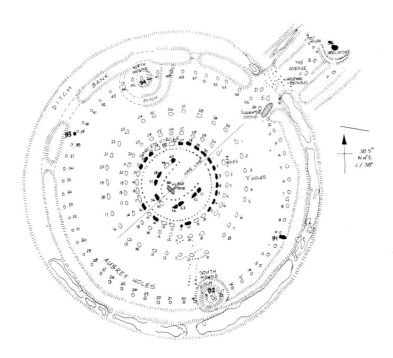

Figure 5.23   Plan of Stonehenge (based on original © HMSO)

  (iv)   Position 93 seen from hole *F*
         Position 94 seen from *C*

   (v)   Hole *H* seen from 93
         Hole *G* seen from 94
         The gap between Sarcen Stones 6 and 7 seen from the gap
         between Trilithons 51 and 52

  (vi)   Position 92 seen from 91
         Position 93 seen from 94
         The gap between Sarcen Stones 15 and 16 seen from the gap
         between Trilithons 55 and 56

B.   (i)   Hole *G* seen from 92
           Hole *A* seen from the centre
           Hole *D* seen from the centre

    (ii)   Position 94 seen from 91
           The gap between Sarcen Stones 21 and 22 seen from the gap
           between Trilithons 57 and 58

   (iii)   Hole *F* seen from the centre

    (iv)   Position 93 seen from the centre
           The gap between Sarcen Stones 20 and 21 seen from the gap
           between Trilithons 57 and 58

     (v)   Heelstone seen from 94
           Hole *B* seen from 94

    (vi)   Position 94 seen from hole *D*

   (vii)   Hole *E* seen from 94

  (viii)   No alignment found

    (ix)   Position 92 from 93
           The gap between Sarcen Stones 9 and 10 and the gap between
           the Trilithons 53 and 54

     (x)   No alignment found

    (xi)   Position 91 seen from centre
           The gap between Sarcen Stones 8 and 9 and the gap between
           the Trilithons 53 and 54

   (xii)   No alignment found

Take each one of these in turn, pencil lightly the alignment on the plan
and measure the angle from the east of this alignment. Using the nomograph, or the formula, find the angle from the east for the event as
defined by the declination of the Sun or the Moon and compare the two
angles. Since the declinations change slightly each year, we must take the
values as they were at the times of the construction of Stonehenge.

Thus, for sunrise furthest north of east the declination is $\quad +23.9°$

sunrise due east $\qquad 0°$

sunrise furthest south of east $\qquad -23.9°$

and similarly for the corresponding sunsets.

Also, for full Moon rise when furthest north the

declination is $\; +29° \; (P_1)$

full Moon rise when furthest north $\qquad +18.7 \; (W_1)$

full Moon rise when furthest south $\qquad -29° \; (P_2)$

full Moon rise when furthest south $\qquad -18.7° \; (W_2)$

and similarly for the corresponding Moon sets.

The declination of the Moon at its equinox can be any angle between $+5.15°$ and $-5.15°$.

If an alignment appears to be blocked by one or more of the trilithons we must remember that Stonehenge was built in stages, the horseshoe of trilithons being one of the later features.

The latitude of Stonehenge is $51.17°$ N.

Note:
Stonehenge Decoded, Gerald Hawkins, Fontana Books 1970 – for further suggested reading.

6

Old astronomical instruments

6.1 THE NOCTURNAL

A sundial, together with a knowledge of the equation of time, can be used for telling the time in the day provided the Sun is shining sufficiently strongly to cast a shadow.

The nocturnal, which was in use from the sixteenth century to about the eighteenth century, can be used for telling the time at night provided that the pole star, Polaris, and the pointers in the constellation of Ursa Major (the Plough or the Big Dipper) can be seen.

The Earth rotates on its axis once in approximately 24 hours. This axis passes close to the pole star, and so the sky seems to rotate uniformly approximately one revolution around the pole star in a day. The line joining the stars Polaris and the pointers of the Plough can therefore be regarded as the hour hand of a 24-hour clock centred on the pole star. Thus if one had a reference position for the line and the corresponding time, one could deduce the time on any other occasion by observing the Polaris–pointers line relative to the reference line.

There is, however, a slight complication in that the Earth is moving in an orbit round the Sun. The sky is therefore displaced relative to the reference line about one degree each day. Some device has to be built into the nocturnal to make allowance for this.

Project 34 shows how a nocturnal can easily be made from card or suitably thin wood.

6.2 PROJECT 34

To construct a simple nocturnal, and to use it to tell the time at night.

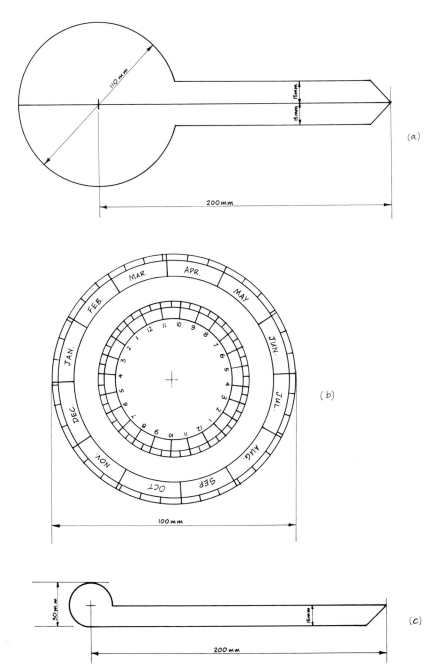

Figure 6.1   Parts of a nocturnal

Figure 6.1 continued

Figure 6.1 shows the three main parts of the nocturnal and suitable dimensions. Figure 6.1(a) represents the back of the nocturnal on which is scribed a reference line. Figure 6.1(b) is the main disc. The outer circle is divided accurately into months, and each month has 10 day subdivisions. The inner circle is divided accurately into 24 equal hours, in this case labelled as two 12 hours. These markings must be arranged so that the radius through March 7 passes through one of the 12 hour markings (see Appendix A9). Figure 6.1(c) is simply a straight edge for lining up the star Polaris and the pointers. The parts are assembled in the order given above and held freely but firmly by a centre pin, which preferably should have a hole through it.

To use the nocturnal, hold the back so that the line scribed on it is vertical and Polaris can be seen through the central eyehole. Turn the disc so that the date lines up with the reference line at the top. Turn the straight edge so that it is in line with the pointers of the Plough. Read the time at the point where the straight edge cuts the hour circle. With the divisions as shown it should be possible to tell the time to about a quarter of an hour. Figure 6.2 shows a typical situation.

6.3   THE ASTROLABE

An astrolabe is a device, invented as long ago as about 240 B.C., for taking altitudes of heavenly or earthly bodies. It is, however, much more than this. Used properly it will, without difficulty, show when and where the Sun or particular stars or planets will rise or set; tell the time of day; convert equatorial co-ordinates into altitude–azimuth co-ordinates and

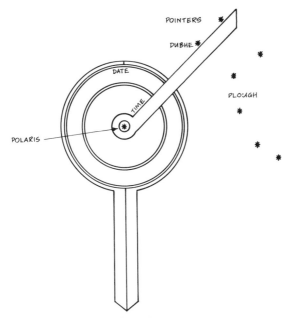

Figure 6.2   Typical configuration

vice versa; give the altitude and bearing of particular stars at any specified time, and so on. It is as accurate as it can be made.

The simple astrolabe to be described here ignores the altitude-taking function and emphasises the uses of the astrolabe as an astronomy 'slide rule'.

The parts of an astrolable are: (1) the mater, with markings on both sides (Figure 6.3 and Figure 6.4); (2) the plate (Figure 6.5), which will be correct for one latitude only. If the astrolabe is to be used in another latitude then another suitable plate will have to be substituted; (3) the rete (Figure 6.6); and (4) two pointers – one for the front of the astrolabe (Figure 6.7(a)) and one for the back of the astrolabe (Figure 6.7(b)).

The markings on the mater represent the zodiac (outer circle) and calendar year (inner circle) on the back (Figure 6.4) and Greenwich Mean Time on the front (Figure 6.3).

The markings on the plate represent lines of equal altitude and lines of equal azimuth; those on the rete represent lines of equal hour angle and lines of equal declination. The rete also carries the positions of certain prominent stars correctly located with respect to the lines of hour angle

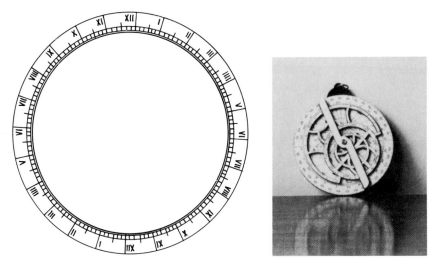

Figure 6.3   (a) The mater (front). (b) The astrolabe front

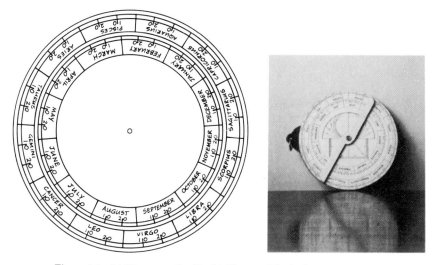

Figure 6.4   (a) The mater (back). (b) The astrolabe back

and declination. Between them they form a flat map of the heavens which is produced by south polar stereographic projection on to the equatorial plane (see Appendix A10). One fundamental property of this map projection is that any circle on the base celestial sphere projects into a circle on the equatorial plane, although, in general, concentric circles do not project into concentric circles.

The markings on the front pointer are those of declination. There are no markings on the back pointer which is used simply as a lining-up device.

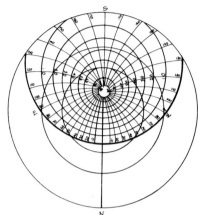

Figure 6.5 The plate

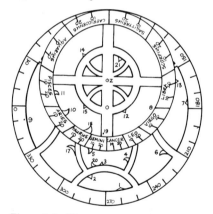

Figure 6.6 The rete

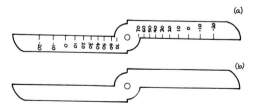

Figure 6.7 The pointers

6.4   PROJECT 35

To construct a simplified astrolabe.

The astrolabe can be made of any suitable, workable sheet material, but thin white card is recommended initially with all the markings made in pencil. The size of the astrolabe will depend on the radius r chosen for the base sphere. We shall take this as 50 mm. The markings on the plate will depend on the latitude φ of the observer. For the sake of example we shall take φ as 52°N but the reader should take his or her own latitude.

(a)   The front of the mater

Copy Figure 6.3, making the radius of the inside circle the same as that of the outside of the plate, namely $r_2 = r \tan 56.75°$, and that of the outer circle about 10 mm larger. Cut round both the inside and outside perimeters to form an annulus with the time scale on it.

(b)   The back of the mater

Copy Figure 6.4, making the outside radius equal to the outside radius of the annulus above. Make the zodiac about 10 mm wide, each sign of the zodiac occupying 30°. Make the calendar also about 10 mm wide, being careful to line up March 21 with Aries 0°. Cut around the outside perimeter only, and glue the annulus above on the reverse side, so that the time scale is shown. This will leave a hollow circular cavity into which the relevant plate (according to latitude) can be placed.

(c)   The plate

Calculate the radius of the Tropic of Cancer $(r_1 = r \tan 33.75°)$, the radius of the equator $(= r)$ and that of the Tropic of Capricorn $(r_2 = r \tan 56.25°)$. With centre C draw the three circles with these radii. The last will represent the perimeter of the plate. Draw NCE horizontally to represent the x axis and a line vertically through C to represent the y axis, the origin being at C.

Calculate the radius R of each circle of altitude, for altitudes a from 0° to 90° in intervals of 10°, from $R = r \cos a/(\sin \varphi + \sin a)$. Calculate the x, y co-ordinates of the centres of these circles from

$$x_a = r \cos \varphi/(\sin \varphi + \sin a), y_a = 0$$

Draw these circles of constant altitude. The observer's zenith z will be the circle of zero radius corresponding to $a = 90°$. The observer's horizon will be the circle corresponding to $a = 0°$.

Calculate the co-ordinates of the centres of the circles of constant azimuth A from $0°$ to $180°$ in intervals of $10°$ from $x_A = -r \tan \varphi$, $y_A = -r.\cot A.\sec\varphi$. With each of these centres C' and radius $C'Z$ draw arcs from the horizon to Z. This completes the plate which should now be cut out along the perimeter and inserted into the recess in the mater so that both N and S are opposite 0 hours on the annulus, but at different ends of the diameter.

(d)   The rete

Draw a circle with centre O and radius equal to that of the plate. Mark around the perimeter of this circle angles from $0°$ to $360°$ in intervals of $10°$ as shown in Figure 6.6. These angles represent hour angles. Calculate $\frac{1}{2}(r_2 - r_1)$. Mark point Z such that $OZ = \frac{1}{2}(r_2 - r_1)$ along the radius from O to the $90°$ mark on the perimeter of the rete. With centre Z and radius $\frac{1}{2}(r_2 + r_1)$ draw a circle which will represent the ecliptic. Draw circles each 5 mm less than the one last drawn. Mark the signs of the zodiac on the ecliptic at $30°$ intervals making sure that Aries $0°$ lines up with $0°$ right ascension. The graduations and divisions between the zodiacal constellations are radial *from* O.

We must now mark on the rete the positions of some prominent stars. For this we shall need the declination δ in addition to the hour angle H for each star. Calculate $r_\delta = r.\tan(45° - \delta/2)$ for values of $\delta = -20°$ to $+70°$ in $10°$ intervals. On a piece of tracing paper draw concentric circles with these calculated radii as in Figure 6.8. Draw also radii of hour angles H at $30°$ intervals on the same piece of tracing paper.

To determine the position on the rete of a star of equatorial co-ordinates (H, δ) fit the tracing of the declination circles over the rete so that their centres and hour angles coincide. Prick through (H, δ) and label this star on the rete. Do this for the stars listed in Table 6.1. The remaining part of the rete is simply tracery to hold the stars in their correct position. The tracery can be copied roughly from Figure 6.6. Cut out the panels between the tracery, avoiding cutting off the star holders just drawn. Then cut round the perimeter of the rete and those points in the centre of the rete which are neither tracery itself nor star holders.

TABLE 6.1

No	Name of star		Hour angle*	Declination
1	Sirius	α Canis Majoris	259°	−17
2	Rigel	β Orionis	282	−8
3	Betelgeuse	α Orionis	272	+7
4	Procyon	α Canis Minoris	246	+5
5	Markab	α Pegasi	14	+15
6	Alphard	α Hydrae	218	−9
7	Deneb	α Cygni	50	+45
8	Denebola	β Leonis	183	+15
9	Diphda	β Ceti	349	−18
10	Hamal	α Arietis	329	+23
11	Regulus	α Leonis	208	+12
12	Dubhe	α Ursae Majoris	195	+62
13	Spica	α Virginis	159	−11
14	Ras Alhague	α Ophiuchi	97	+13
15	Alpheratz	α Andromedae	358	+29
16	Altair	α Aquilae	63	+9
17	Aldebaran	α Tauri	291	+16
18	Schedar	α Cassiopeiae	350	+56
19	Capella	α Aurigae	281	+46
20	Mirfak	α Persei	310	+50

* See Appendix A.12

(e) The pointers

Shape two pointers as in Figure 6.7. The radius of each pointer will be the same as the outside radius of the mater. Fit the declination tracing paper over one pointer so that their centres coincide and prick through the tracing paper, at the edges of the pointer, the declination marks.

(f) Assembly

Punch a hole about the size of a 6 BA screw in the centre of all the parts. Assemble the parts on to a 6 BA screw in the order pointer (a), mater, plate, rete and finally pointer (b) with a washer underneath the head of the screw and one under the nut.

Values of r_1, r_2, x_a, y_a, x_A, y_A and r_δ for the chosen value of $r = 50$ mm   and the chosen value of $\varphi = 52°$ are given in Appendix A13.

To use the astrolabe, say to find the time of sunrise on May 24:

(a)   turn the astrolabe so that the blank pointer is uppermost;

(b)   set one radial edge of the pointer opposite May 24;

(c)   read the corresponding zodiacal sign and angle (Gemini 3°);

(d)   turn over the astrolabe so that the declination pointer is uppermost;

(e)   rotate the rete until Gemini 3° on the ecliptic circle just lies on the horizon $a = 0$ (near E for sunrise);

(f)   set one radial edge of the pointer opposite Gemini 3° and read GMT on the time scale on the mater (= 4 h GMT).   Add 1 hour if BST is operating   (= 5h BST).

Thus the sun rises at 5 o'clock BST on May 24.

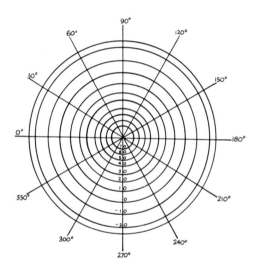

Figure 6.8   Declination circles

6.5   A LINEAR ASTROLABE

The projection used for the plate and rete of the circular astrolabe described in Project 35 was south polar stereographic onto the equatorial plane. If instead we use Gall's stereographic onto a cylinder we can unroll the cylinder to obtain a rectangular map of the sky. The lines of equal altitude and the lines of equal azimuth form more complex curves on the plate than for the circular astrolabe but the rete is a rectangular grid, with equal spacings for equal increments of hour angle but with unequal spacings for declination. When assembled as shown in Figure 6.9 the rete and plate, associated with the other markings of the zodiac,

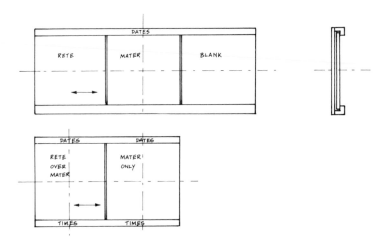

Figure 6.9   Assembled astrolabe

calendar date and time, can be operated by a linear motion in the manner of a slide rule.

6.6   PROJECT 36

To make a simple linear astrolabe.

Cut a piece of thick card to dimensions 376.8 mm wide and 116.6 mm high. Divide the card into three equal parts by drawing vertical lines 125.6 mm apart. In the centre section draw horizontal lines 10 mm from the top and bottom respectively, leaving 96.6 mm between them. Draw vertical and horizontal lines to divide this central area into four equal parts each 31.4 mm wide by 48.3 mm high. Take the origin as the left hand end of this horizontal line in the centre section (see Figure 6.10). From Table A2 (Appendix A14) plot lines of equal altitude (for latitude 52°N) to a horizontal scale   2 mm = 1 unit   in the Table and a vertical scale   2 mm = 1 unit.   From Table A3 (Appendix A14) plot lines of equal azimuth to the same scales. Construct on another piece of thick card 125.6 mm long and 10 mm wide a time scale as shown in Figure 6.10 and on a similar card construct a calendar scale making sure that

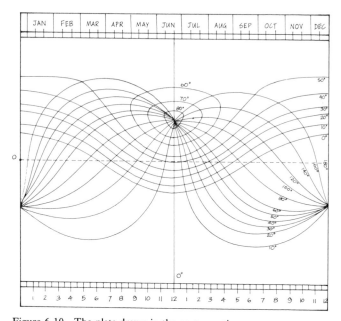

Figure 6.10   The plate drawn in the centre section

March 21 is in the position corresponding to 6 hours on the first scale. Stick these scales at the bottom and top of the central portion of the plate just constructed.

On another piece of card or paper which will just fit the central recess 125.6 mm wide and 96.6 mm high, construct the rectangular grid of the rete as shown in Figure 6.11. The hour angles are spaced 10.46 mm for

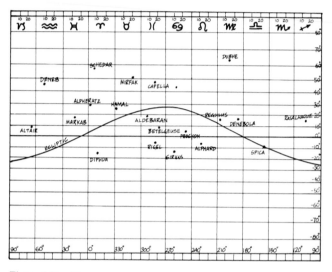

Figure 6.11   The rete

each 30°. The distances from zero declination to the other declination lines are:

TABLE 6.2

0 – 10°	4.22 mm
0 – 20	8.51
0 – 30	12.94
0 – 40	17.57
0 – 50	22.51
0 – 60	27.87
0 – 70	33.81
0 – 80	40.51
0 – 90	48.28

Next plot the ecliptic on the rete from the co-ordinates given below:

TABLE 6.3

Hour angle	Declination
78°.4	−23°.0
46 .3	−17 .5
18 .3	−7 .5
350 .7	+4 .0
321 .7	+15 .0
291 .0	+22 .0
260 .0	+23 .4
228 .7	+18 .0
199 .7	+7 .5
184 .0	+3 .0
143 .7	−14 .3
114 .0	−21 .6

Also plot the positions of some of the stars given in Table 6.1 in Project 35 and mark in the constellations of the zodiac as in Figure 6.11.

The rete must now be made transparent. This may be done by passing the drawing through a machine which makes overhead projector transparencies.

As a check we may find the time deduced by the Host in *The Canterbury Tales* in the Prologue to 'The Man of Laws Tale' for a date of April 18 when the Sun's altitude was 45°. This should be 10 o'clock.

Set the rete centrally and read the sign and degrees of the zodiac opposite April 18 (Aries 29°). Now slide the rete sideways to the right until the point on the ecliptic directly beneath Aries 29° lies on the 45° altitude curve on the plate below. Read the time directly below this point.

7

Stars

7.1 STAR CHARACTERISTICS

When we look at the sky we may be forgiven if, at first, we do not notice the wide variation in the appearance of the stars. Not only do the stars differ in brightness, but the stars are not all of the same colour. For instance, the star Rigel, in one corner of the constellation of Orion, appears to be brighter than the star Betelgeuse in the opposite corner of the same constellation. Rigel is a bluish-white star, while Betelgeuse is red.

Less immediately obvious is the fact that some of the stars do not shine with the same brightness at all times. These are the variable stars, of which there are several types.

The brightness of some stars, the irregular variables, fluctuates in an irregular manner. The period of fluctuation of brightness may be as long as several hundred days, and in the case of these long period variations both the period and the greatest brightness attained are usually variable.

For some stars the pattern of change in brightness repeats itself continually over equal periods of time. Stars in this category include the Cepheid variables, with typical periods between one and fifty days, but mostly about five days. The cluster-type Cepheids, so called because they are found mostly in globular clusters, have a period of less than one day. The Cepheids are important because they help in distance determination.

Another group of the so-called variables are the spectroscopic binaries. Here there are two stars revolving about their common centre of mass, but they may be so close together that neither the naked eye nor photography can resolve the components and they appear to be a single star. When the stars are side by side as viewed from the Earth, the total radiation received by the observer will be the sum of the individual radiations of the two stars. If, however, star A is in front of star B, then there will be some reduction in radiation received by the observer, since star A blocks out some of the radiation of star B and prevents it from reaching the Earth. Thus a periodic variation in brightness is observed.

Then there are the novae and supernovae. These are stars which suddenly increase in brightness tremendously in a short time and then fade again. Supernovae are relatively rare.

Finally we have the pulsars and quasars, but these are not naked eye objects. However, we shall say more about them later.

7.2 THE NATURE OF A STAR

So far we have looked at stars from the standpoint of radiation received, either constant or variable. Indeed, all the information we have about stars in general is carried in this radiation.

It is, however, pertinent to ask what a star is, what is its radius, its mass, its temperature and, in view of the amount of radiation from a star such as the Sun, how the temperature is maintained not only on the surface of the star, but also within the star. The surface temperature is the only one which we can observe directly since it is associated with the colour of the star.

It is only within the years of this century that a real understanding of stellar structure has been developed. A star is a radiating sphere of gas composed mainly of hydrogen and helium, but also with some heavier elements. It maintains its equilibrium by balancing the gravitational attraction of the matter of which the star is composed with the pressure created by this attraction and by radiation. The radiation is maintained by nuclear reactions within the star due to the very high internal temperatures, especially near the core, where, in the Sun, the temperature is about 15 million Kelvin.

When we think of a gas, we normally think of a very tenuous substance such as the air in our own atmosphere. The density of such a gas, that is its mass per unit volume, is very small, and in the outer layers of the stars the density of its gas is indeed small. In the innermost parts of the stars however, the density is much greater and may exceed 100 times that of water. It is mainly due to the stripping of the electrons from the atoms of a star that materials of such density can be regarded as behaving like a perfect gas. A perfect gas is one which obeys a certain law relating the pressure, temperature and density. Those readers who are familiar with elementary physics will have a knowledge of this law, if only in the two forms stated by Boyle and Charles.

In the projects which follow, we shall investigate in part this complicated subject, and show by graphical means how to determine the mass and the pressure distribution within a star, as well as the properties associated with the brightness of stars. At the same time it should be realised that some properties, such as the distribution of density within a star, do not lend themselves readily to simple treatment, and we shall have to take some of these, so to speak, on trust.

7.3 THE MASS OF A STAR

The density of a star is not uniform throughout, being very dense at the centre and very much less dense in the outside layers. We shall assume that the density is known at any given distance from the centre of the star.

Consider a very thin-walled sphere of radius r within the star, of wall thickness t. We can say that the volume of matter making up this thin sphere is given by the product of the surface area and the wall thickness, namely $4\pi r^2.t$.

If ρ is the density of the matter making up the sphere, regarded as constant for such a small variation of radius, then the mass of the thin sphere will be $4\pi r^2 t\rho$. If we now work out the quantity $4\pi r^2\rho$ and plot this as ordinate against r as abscissa, we shall obtain a curve such as the one shown in Figure 7.1.

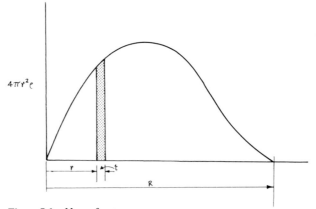

Figure 7.1 Mass of a star

It will be seen from this that the mass of the thin sphere of radius r is equal to the area of the shaded strip. If we now add up all similar strips from the value $r = 0$ to the value $r = R$, the radius of the star, we see that the total mass of the star is represented by the area under the whole curve.

7.4 PROJECT 37

Given the density of a star at points along a radius, to determine the mass of the star contained within each given radius, and to arrive at a close approximation to the mass of the whole star.

Data

The star to be considered is our own Sun, which may be regarded as an average star.

As quoted before, the mass of the Sun is about 2×10^{30} kg and the radius of the Sun is 696 000 metres.

If we denote the radius of the Sun by R and the radius of any sphere within the star by r, then the density ρ at different radii (see *Outline of Astronomy* Vol II, H. H. Voigt, p. 310, Sijthoff and Noordhoff International Publishers bv) is given in the Table below:

TABLE 7.1

r(km)	ρ(kg m^{-3})	$4\pi r^2\rho$(kg m^{-1})
0	134×10^3	
28 000 (0.04R)	121×10^3	
70 000 (0.1R)	86×10^3	
139 000 (0.2R)	36×10^3	
209 000 (0.3R)	13×10^3	
279 000 (0.4R)	4.1×10^3	
348 000 (0.5R)	1.3×10^3	
418 000 (0.6R)	0.40×10^3	
488 000 (0.7R)	0.12×10^3	
556 000 (0.8R)	0.04×10^3	
627 000 (0.9R)	0.01×10^3	
696 000 (R)	virtually zero	

Complete the column headed $4\pi r^2 \rho$.

Plot the figures in this column as ordinate using a scale of 20 mm $= 1 \times 10^{21}$ kg m^{-1} against the figures in the first column (r) as abscissa using a scale of 20 mm $= 1 \times 10^8$ m.

Determine the areas, by counting squares, under the graph from $r = 0$ to each value of r given in the first column except $r = 28\,000$ km. These values will be needed in the next Project.

We now convert the number of squares under the whole graph to kg units. Each square 10 mm $\times$ 10 mm represents $(0.5 \times 10^8) \times (0.5 \times 10^{21}) = 0.25 \times 10^{29}$ kg and so 100 small squares, each 1 mm $\times$ 1 mm represents 0.25×10^{29} kg.

Using this information, determine the mass represented by the area under the whole curve and compare this with the mass of the Sun given above under Data.

7.5   THE PRESSURE INSIDE A STAR

Most stars remain the same over long periods of time, and it is reasonable to suppose that the material in the star is neither being pushed out due to the buoyancy of the material below it, nor pulled in due to the gravitational attraction between the inner material and the outer material.

If we now consider the forces acting on our thin sphere of radius r and thickness t, we have:
(a) An inwards gravitational force due to the attraction of the 'solid' sphere of gas on the inside of the thin sphere, on that thin sphere. In this case all the mass of the thin sphere is concentrated at radius r, and all the mass of the 'solid' sphere can, for this purpose, be regarded as being concentrated at its own centre (see, for example, *Mechanics of Particles and Rigid Bodies*, by J. Prescott, third edition, p. 220, published by Longmans, Green and Company).

Thus, using Newton's law of gravitation (see Section 2.11), this force is given by

$$F = \frac{Gmm'}{r^2}$$

where $m = 4\pi r^2 t \rho =$ mass of the thin shell
$m' =$ mass within the shell.

Hence the inwards gravitational force on the thin shell is given by

$$F = \frac{G.4\pi r^2 t\rho.m'}{r^2} = G.4\pi t\rho.m' \qquad [\text{i}]$$

(b) An outwards force due to the excess pressure of the material inside the sphere as compared with the pressure of the material outside the thin sphere. If this excess pressure is p, then the net outwards force is given by

$$F = p.4\pi r^2 \qquad [\text{ii}]$$

since the surface area of the thin sphere on which the pressure acts is $4\pi r^2$.

(c) It might be thought that there would be a gravitational attraction on the thin sphere due to the material outside it, but it can be shown that this is zero (see J. Prescott, p. 20 again).

Thus, for balance of the forces within the star,

$$p.4\pi r^2 = G.4\pi t\rho.m'$$

or, pressure *difference* at radius r is

$$p = \frac{G.\rho m'.t}{r^2}$$

Hence, if we require the pressure at radius r we sum up the pressure differences from the outside radius, where the pressure is zero, to the radius at which we require the total pressure. To do this, we may plot a graph of $G.\rho.m'/r^2$ against r. The area under this graph, in the correct units, from $r = R$ to $r = r$ will give the pressure at radius r. As with Project 37 on the determination of the mass it will be realised that both ρ and m' will depend on the radius. These values have been determined in the previous exercise. The area concerned is shown in Figure 7.2.

7.6 PROJECT 38

To determine the pressure at a specific radius in a star given the density at a few values of the radius.

Data

The star chosen is the same as that for Project 37, namely the Sun.

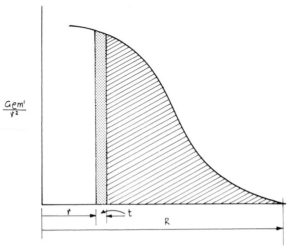

Figure 7.2   Pressure inside a star

Mass of Sun $= 2 \times 10^{30}$ kg
Radius of Sun $= 696\,000$ km
Gravitation constant $G = 6.67 \times 10^{-11}$ kg^{-1}m^{3}s^{-2}

TABLE 7.2

r(km)	ρ(kg m^{-3})	m'(kg)	$G\rho m'/r^2$ (kg m^{-2} s^{-2})
0	134×10^3	0	
69 600	86×10^3	0.142×10^{30}	
139 000	36×10^3	0.664×10^{30}	
209 000	13×10^3	1.215×10^{30}	
279 000	4.1×10^3	1.597×10^{30}	
348 000	1.3×10^3	1.798×10^{30}	
418 000	0.40×10^3	1.891×10^{30}	
488 000	0.12×10^3	1.932×10^{30}	
556 000	0.04×10^3	1.949×10^{30}	
627 000	0.01×10^3	1.954×10^{30}	
696 000	0	1.956×10^{30}	

The values for the density ρ are those previously assumed, and the values of the mass m' contained in a sphere of radius r are those obtained graphically in Project 37. The values of m' have been given for readers who have not carried out this project. For readers who have, the values obtained can be used.

Complete the last column using the quantities given or obtained.

Plot the values of $G\rho m'/r^2$ as ordinate using a scale of $10 \text{ mm} = 0.1 \times 10^8 (\text{kgm}^{-2}\text{s}^{-2})$ against the values of r (km) using a scale of $20 \text{ mm} = \times 10^8 (\text{m})$.

Working from the value of $r = 696\,000 \text{ km}$ find, by counting squares or otherwise, the area under the curve to $r = 69\,600 \text{ km}$.

For the scales used, a square
$$1 \text{ mm} \times 1 \text{ mm} = (0.01 \times 10^8) \times (0.05 \times 10^8)$$
$$= 5 \times 10^{12} \text{ Nm}^{-2}.$$

Using this information determine the pressure at a radius of $69\,600$ km from the centre of the Sun and compare this with that given in Voigt, namely $1.37 \times 10^{16} \text{ Nm}^{-2}$.

7.7   THE BRIGHTNESS OF THE STARS

The brightness of a star is the total visible light which it emits in a given time. The stars radiate light energy in all directions, and we receive only a very small part of the total from each. The total energy can be thought of as passing through a series of spherical surfaces, each with the star as centre. The surface areas of these spheres are proportional to the squares of their respective radii. Thus the radiation falling on say one square metre of a sphere of radius r will carry on, in a cone, and pass through an area of four square metres of a sphere of radius $2r$. Put in another way, the radiation falling on one square metre of the outer sphere will be only one quarter of that falling on one square metre of the inner sphere. The brightness therefore falls off inversely as the square of the distance of the observer from that star.

Thus, of two equally bright stars, that is, of two stars emitting the same amount of visible light in the same time, the one which is the nearer will appear to be the brighter. We refer to the brightnesses of stars as we see them as the *apparent* brightnesses.

To compare the actual brightnesses of stars we need to cancel out the effect of distance. We therefore calculate from the apparent brightnesses and the distances of the stars, what the brightnesses would be at a standard distance. We use the inverse square law, the standard distance chosen being 10 parsecs. The parsec is a convenient unit of distance when dealing with the stars. Let us imagine that the Earth's orbit is turned so that the major axis is at right angles to the line joining the

Earth to the star. The angle between this line and a line from one end of the major axis to the star, expressed in seconds of arc, is a measure of the distance of the star. If the angle is one second of arc, the distance is called one parsec. If the angle were twice as large, that is two seconds of arc, the star would clearly be nearer and its distance equal to one half of a parsec, and so on. The angle itself is referred to as the parallax of the star. A parsec is about 30.9×10^{12} km or in a more popular unit, 3.26 light-years.

The brightnesses at the standard distance of 10 parsecs are known as the *intrinsic* brightnesses.

7.8   THE MAGNITUDE OF THE STARS

When we refer to the magnitude of a star we are referring to its apparent brightness on a scale which will now be described, and not to its physical size.

If the apparent brightnesses of the two stars are B_1 and B_2 respectively, then the difference in the magnitudes m_1 and m_2 of the stars is given by the relation

$$m_2 - m_1 = 2.5 \log_{10} \frac{B_1}{B_2}$$

If we assign a value of magnitude to a particular star of known brightness to act as a datum, then the magnitude of any other star of known brightness can be calculated. The standard star used to be the pole star, Polaris, which was allocated a magnitude of 2.12, but it was found that the brightness of Polaris varies. Today a combination of standard stars is used.

It will be found on this scale that most of the brightest stars visible to the naked eye have magnitudes about 1, while those just visible to the naked eye have magnitudes round about 6. In fact the origin of this method of specifying brightness lies in the approximate division of the visible stars into six classes according to the estimates of the naked eye.

7.9   RESPONSE OF THE EYE TO BRIGHTNESS

If we are measuring some physical quantity with an instrument which does not respond linearly, then uniform increments in input will not show up as uniform increments on the scale of the measuring instrument. Such is the case with the eye, since to see uniform steps in apparent brightness, the brightness of the object observed must change in a logarithmic manner.

This is taken care of in the astronomical definition of the scale of magnitude m of stars given above, viz

$$m_2-m_1 = 2.5 \log_{10}(B_1/B_2)$$

where linear increments in magnitude, as observed by the eye, are proportional to the logarithm of the ratio of the corresponding brightnesses B, which is, in fact, why the eye can accommodate such large differences in brightness. Thus a factor of 100 in real brightness produces a change of only 5 magnitudes.

7.10 DEMONSTRATION 5

An effective demonstration of the relationship between brightness and magnitude can be made using two white discs and two small electric motors or other devices for rotating the discs. The size and material of the discs are in no way critical. Those used by the author are made of cardboard and are each 300 mm diameter. The connection between disc and motor shaft will depend upon the resources available. One simple method is to make up a collar as shown in Figure 7.3.

Each disc is divided into 6 rings of equal width radially. The outer ring is left completely white. The other rings are partially blocked in with black paint or ink or more easily black card stuck on so that as we move towards the centre, progressively larger proportions of the rings are blocked out. Thus when the discs are spun, the differences in the brightnesses of the rings can be seen, going from the white one at the perimeter to much darker at the centre.

Since the uninked area of any ring relative to the total area of that ring can be taken as a measure of true brightness we can work out the angular width of the uninked portion of each of the rings on the first disc so that the true brightness proceeds in linear stages from the centre outwards, and similarly for the second disc so that the true brightness proceeds in a logarithmic manner from the centre outwards.

These are for disc 1, Figure 7.4(a): 15, 30, 45, 60, 75, 90 degrees
        for disc 2, Figure 7.4(b): 15, 21, 31, 44, 63, 90 degrees
Each successive ring in disc 2 has relatively 43 per cent more white than the previous ring.

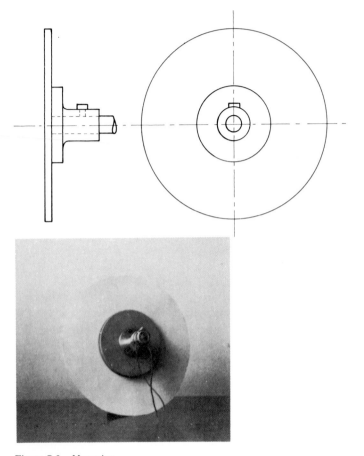

Figure 7.3   Mounting

Observation of the spinning discs should show non-uniform steps in apparent brightness for the linear disc and equal steps for the logarithmic disc.

For the method of calculating the angles for the logarithmic disc the reader is referred to Appendix A15.

7.11   DETERMINATION OF STELLAR MAGNITUDES BY PHOTOGRAPHY

One of the basic measurements in astronomy is the determination of the magnitude of astronomical objects. One widely used method is to take a

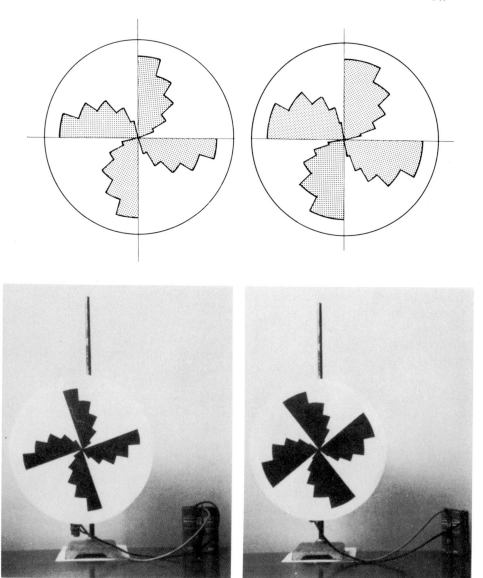

Figure 7.4 (a) Linear disc. (b) Logarithmic disc

photograph of the object through a telescope. A photographic plate has an advantage in that it can record the radiation falling upon it over the time of the exposure. It adds light to light, whereas the eye can only see the light at a particular instant. Thus a photographic plate can record images of faint stars which cannot be seen with the eye even through a telescope. Another advantage of a photographic plate is that it can record the images of many stars at the same time.

The more light which falls on the photographic plate the more is the degree of blackening (on the negative). The final size of each image also depends upon the properties of the photographic emulsion, so that the distribution of image diameters does not give the distribution of brightness of the stars. The size of a stellar image is not a measure of the physical size of the star.

Further, because the Earth's atmosphere is not at rest, the position of the image of the star on the plate wanders slightly, so that the centre of the image will be blacker than at points further from the centre.

It is clear then that calibration of each photographic plate is necessary to determine a relationship between magnitude and image diameter so that magnitude can be attributed to the stars on the plate.

Project 39 is designed to try to find this relationship for a particular photographic plate.

7.12   PROJECT 39

(i) To determine a relationship connecting the apparent magnitude of stars on a photographic plate with the diameters of their images on the plate, that is, to calibrate the plate.
(ii) To deduce the magnitudes of three stars on the same plate using the calibration curve.

Data

The provision of an actual photographic plate in this situation is not usually possible, and so measurements will be taken from a photographic print (Figure 7.5). This shows an area of the sky near to the North America nebula, so called because of its remarkable resemblance in outline to the North America continent. It is a negative print, that is, bright star images appear black and dark areas appear bright. Such a print allows more accurate measurements to be made.

The Table below gives the apparent magnitudes of a number of stars on the photograph supplied. A second print is supplied (Figure 7.6) so that stars can be located on the photograph.

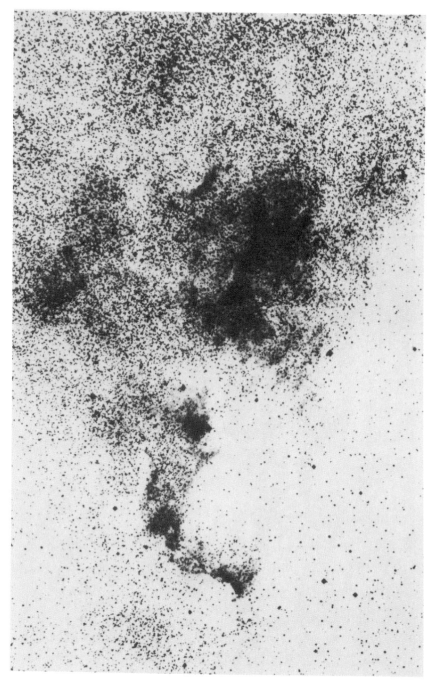

Figure 7.5   North American nebula

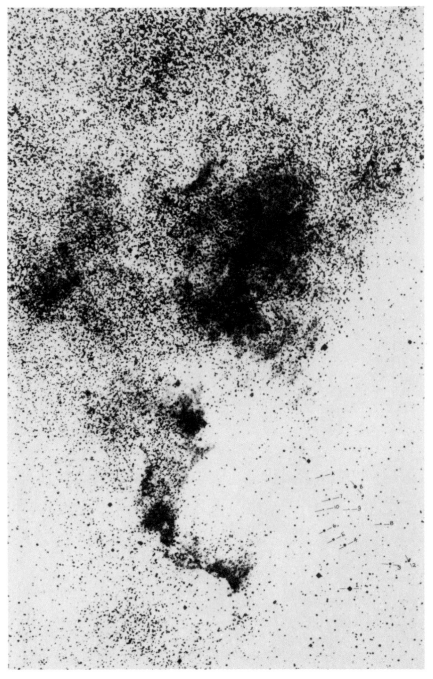

Figure 7.6 Location of stars for calibration

TABLE 7.3

Star number	Apparent photographic magnitude
1	7.9
2	9.7
3	9.4
4	8.5
5	9.4
6	8.9
7	10.5
8	12.2
9	14.8
10	16.2

A magnifying device is required for this project, and also an accurate scale with a bevel edge marked preferably in millimetres. The procedure to be followed is similar to that used by the professional astronomer, who would use a more elegant and accurate piece of equipment, such as an iris diaphragm photometer. For our purpose, if a suitable miscroscope or eye-piece fitted with a graticule is available, well and good, but failing these, tolerable results can be obtained by using the centre of the field of view of a magnifying glass.

Select a star from the foregoing list and measure its diameter on the photograph. In doing so, see that the star image is in the centre of your magnifying device, and that the bevel edge of your scale is tilted so that this edge is in contact with the photograph.

Repeat this procedure for all the stars whose photographic magnitudes are given in the Table.

Plot the measured diameters as abscissae against magnitudes as ordinates. It is usually found that the calibration curve can be expressed in a form according to one of the three equations given below:

$$m = a - b.d$$

$$m = a - b.d^{\frac{1}{2}}$$

$$m = a - b.\log_{10} d$$

In the first case, if we plot diameter d against magnitude m we shall find the points lying on a straight line if that law is obeyed. Similarly, a straight line will be obtained in the second case if we plot $d^{\frac{1}{2}}$ against m, if that law is obeyed. In the third case, we obtain a straight line graph if we plot $\log_{10} d$ against m, provided that this is the law followed by the two variables d and m.

In all three cases a scale for m of   10 mm = 1 magnitude   is suggested. For the first case, a diameter of image of 0.1 mm should be represented on the graph by 20 mm. Put the origin in the left hand corner of the graph sheet.

For the second case, a value of $d^{\frac{1}{4}}$ of 0.1 (d in mm) should be represented by 20 mm. Put the origin in the bottom left hand corner as before.

For the third case, the value of $\log_{10} d$ of 0.1 (d in mm) should be represented on the graph by 20 mm. Put the origin at the bottom of the graph sheet, but at least 170 mm from the left edge of the graph sheet.

Select the graph which gives the best straight line fit to the points, and read off from it the magnitudes of the two stars A and B corresponding to their measured diameters. Check these with the values 11.8 and 7.3 respectively.

7.13   RELATION BETWEEN APPARENT MAGNITUDE AND ABSOLUTE MAGNITUDE

We have said that the brightness which a star would have if placed at a standard distance of 10 parsecs is known as its intrinsic brightness. In a similar way, the magnitude which a star would have at a distance of 10 parsecs is known as its absolute magnitude.

Let the apparent magnitude of a star $= m$
the parallax of the star $= p''$
the absolute magnitude of the star
  (at a distance of 10 parsecs) $= M$
the apparent brightness of the star $= b$
the intrinsic brightness of the star $= B$
Then the distance of the star $= 1/p$ parsecs

Since brightness varies inversely as the square of the distance,

$$\frac{b}{B} = \frac{10^2}{(1/p)^2} = 100p^2$$

But from the formula relating magnitude and brightness

$$M - m = 2.5 \log_{10} \frac{b}{B}$$

Hence, $M - m = 2.5 \log_{10}(100p^2) = 2.5 (\log_{10} 100 + \log_{10} p^2)$

or, $M - m = 2.5 (2 + 2 \log_{10} p)$

or, $M - m = 5 + 5 \log_{10} p$

Finally $M = m + 5 + 5 \log_{10} p$

7.14   PROJECT 40

(i)   To draw a chart relating apparent magnitude, absolute magnitude and the parallax of stars.

(ii)   To determine from the chart the absolute magnitude of a number of stars.

On a piece of graph paper, set up axes with the origin at its centre with the apparent magnitude m as abscissa and the absolute magnitude M as ordinate. Mark each axis from 0 to $+3$ to a scale of   25 mm = 1 magnitude,   the same scale being used for each axis. Mark also each axis in the negative direction from 0 to -3.

For all stars with a parallax of $1.0''$ (or at a distance of 1 parsec), we have from the magnitude relationship just deduced

$$M = m + 5 + 5 \log_{10} 1.0$$

or, $M = m + 5$ since $\log 10_{10} 1.0 = 0$

This is a straight line at $45°$ to each axis and passing through the point whose co-ordinates are $(-3, 2)$. Draw this line on the graph.

For all stars with a parallax of $0.1''$ (or at a distance of 10 parsecs) we have the relationship

$$M = m + 5 + 5 \log_{10} 0.1$$

or, $M = m + 5 + 5 \times (-1)$

or, $M = m$

This is a straight line through the origin at $45°$ to each axis, and this may be drawn on the graph.

Intermediate lines may be drawn for values of parallax p between $1.0''$ and $0.1''$, and also lines may be drawn down to a parallax of $0.01''$ in a similar manner, by substituting the appropriate value of parallax in the formula and evaluating.

The Table below gives the apparent magnitude and the parallax for four of the brightest stars:

TABLE 7.4

Star	Apparent magnitude	Parallax (")	Absolute magnitude
Sirius	−1.58	0.371	1.3
Procyon	0.48	0.312	3.0
Achernar	0.60	0.049	−0.9
Pollux	1.21	0.101	1.2

For any of these stars, read off the apparent magnitude along the horizontal apparent magnitude scale, move parallel to the absolute magnitude scale until the diagonal parallax line is encountered corresponding to the parallax figure given in the Table above. Then move horizontally to intersect the absolute magnitude scale. Read this value and compare it with the relevant absolute magnitude given in the last column of the Table above.

7.15   NOMOGRAPH ALTERNATIVE

We saw in Section 7.13 that the apparent magnitude m, the absolute magnitude M and the parallax p of a star are connected by the relationship

$$M = m + 5 + 5 \log_{10} p$$

Given any two of the three variables m, M and p, the third can easily be found. Such an equation, while not difficult to solve, lends itself to a nomographic solution which is particularly useful if the equation must be solved repeatedly.

Project 41 describes a suitable nomograph for this purpose. The relevant theory is given in Appendix A6.

7.16   PROJECT 41

To construct a nomograph to represent the relation between the absolute magnitude M of a star, its apparent magnitude m and its parallax p.

Parallel to the longer sides of the paper, draw three lines such that the distance between adjacent lines is 50 mm. Starting at -2 and progressing upwards at a scale of 10 mm = 1 magnitude, construct the uniform scale for apparent magnitude m up to a value of $+15$, on the right hand line. Using a scale of 5 mm = 1 magnitude, construct the uniform scale for absolute magnitude M starting at the bottom with a magnitude of say -14 and ending at the top with a magnitude of $+20$, making sure that this final value of $M = 20$ is in a direct horizontal line with $m = 15$. The M scale should be marked on the centre line.

On the left hand scale, mark $p = 1.00''$ opposite $m = 15$ and $M = 20$, mark $p = 0.1''$ opposite $m = 10$ and $M = 10$, mark $p = 0.01''$ opposite $m = 5$ and $M = 0$ and finally $p = 0.001''$ opposite $m = 0$ and $M = -10$.

For the intermediate graduations on the p scale we proceed as follows, since these graduations have to be on a logarithmic scale. The distance between 0.1 and 1.00 is 50 mm, which therefore represents $\log_{10} 1.0 - \log_{10} 0.1 = 0 - (-1)$. Now $\log_{10} 0.75 = 1.875 = -0.125$. Hence the linear distance on the p scale between 1.0 and 0.75 is 0.125×50 mm = 6.25 mm.

Similarly, for $p = 0.50''$, $\log_{10} 0.5 = \bar{1}.699 = -0.301$. Hence the linear distance on the p scale between $1.0''$ and $0.5''$ is $0.310 \times 50 = 15.05$ mm, and so on for as many markings as we wish.

For the gap between $0.1''$ and $0.01''$ the divisions will appear in the same relative places. For example $0.075''$ will be 6.25 mm from $0.1''$. The same will apply for the interval $0.01''$ to $0.001''$.

To use the nomograph, a straight edge such as a ruler is placed across the nomograph so as to intersect the two given values. The intersection of the ruler with the third scale then gives the corresponding third value required.

As an example, determine the absolute magnitude M of a star whose apparent magnitude is 4 and whose parallax is $0.50''$. Compare this with the value derived from the formula, namely

$$M = m + 5 + 5 \log_{10} p$$

$$M = m + 5 + 5 \log_{10} 0.50$$

$$M = 9 + 5 \times \bar{1}.6990 = 9 - 5 \times 0.301$$

$$M = 9 - 1.50 = 7.50$$

Hence the absolute magnitude of the star = 7.50.

7.17   THE SPECTRUM

When a narrow beam of white light is passed through a transparent prism it emerges as a band of colours which change from red at one end to violet at the other through the colours of the rainbow. This band of colour is called a continuous spectrum.

We get much the same sort of spectrum from the prism if the light entering it comes from a hot, glowing body, irrespective of what the hot body is made. If, however, the hot body is heated until it vaporizes and the light from the hot vapour is passed through the prism the band of colours is weakened but is crossed by a series of bright lines. Each element in the vapour provides one or more of the bright lines, each in a definite position relative to the others. If the vapour is sufficiently hot it is therefore possible to tell by inspection of the spectrum which elements are present in the vapour. The same is true of a glowing rarefied gas. Such a spectrum is known as an emission spectrum.

If the light from a glowing gas or vapour passes through a similar gas at a lower temperature before entering the prism the bright lines are replaced by dark lines. In this case we obtain an absorption spectrum.

The latter is just the situation which exists in stars since the outer layers of the atmosphere are at a low pressure and cooler than the inner layers. Thus we can expect the continuous spectrum of a star to be crossed by a number of dark lines, although some of the hottest stars exhibit emission lines as well.

The stars can be classified according to the spectra which they produce. About 95 per cent of the stars can be put into classes which are labelled O, B, A, F, G, K and M, according to special characteristics of the spectra. For example, in type A stars the lines representing hydrogen are prominent, whereas in type G the lines representing calcium are strong and the hydrogen lines are considerably weaker.

The order of the classes given above is also the order of the surface temperature of the stars, the ones of type O having the highest temperature. It is also generally the order of absolute magnitude of the stars.

Approximate corresponding values of temperature, absolute magnitude and colour of the stellar surfaces in the various classes, are given below.

Our own Sun is a G type star.

It will be clear that the stars do not fall cleanly into the classifications given, and that intermediate graduations are necessary. This is done by

TABLE 7.5

Spectral class	Temperature (K)	Absolute magnitude	Colour
O	35 000	−5	Bluish-white
B	23 000	−2	Bluish-white
A	11 000	1.5	Bluish-white
F	7 500	3.5	Yellow-white
G	5 800	5	Yellow
K	4 600	6.5	Orange
M	3 000	11	Red

attaching a digit from 0 to 9 inclusive to the letter of the classification. Thus the star Spica is classified $B2$, that is between B and A, but nearer to B, while the star Procyon is classified $F5$ or halfway between F and G.

7.18   THE DIFFRACTION GRATING

Light from the stars comes to us in a succession of waves, as in Figure 7.7. The wavelength of the light is also defined in Figure 7.7.

If we have two slits emitting monochromatic light, that is light of one wavelength only, such that the narrow slits A and B are the same distance from an observer O, then the crests and troughs from A and B will reach O at the same time, and we say that the waves are 'in phase'. In this case the illumination at O will be bright due to reinforcement of one wave by the other as in Figure 7.8.

If the distance which the light has to travel from B to a second point P is greater by half the wavelength of the light $(\lambda/2)$ than the distance from A to P, we shall get troughs from A coinciding with crests from B. The waves are completely 'out of phase', and darkness results at P. Similarly, if the distances travelled to a point Q differ by λ, then we shall once more be in phase and brightness ensues. Thus we get a series of dark and bright lines on the line OPQ (Figure 7.9).

The distances OP and PQ will, of course, depend on λ, the wavelength of the light used. Thus for red light Q will be further from O than for violet light because λ_{red} is greater than λ_{violet}. A white light should thus

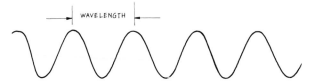

Figure 7.7   Wave nature of light

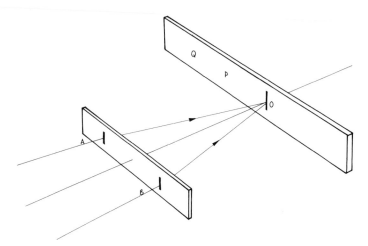

Figure 7.8   Light from two slits

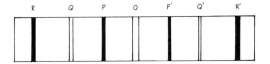

Figure 7.9   Formation of dark and light bands

produce a spectrum in the general region Q. Due to the low number of sources this would be very faint, and we can intensify the spectrum by using many more slits very close together, so that the effects from each slit can add. Such a large number of slits may be obtained from a diffraction grating, on which can be ruled about 600 lines per millimeter (Figure 7.10). The size of the slits on the diffraction grating has to be about the same order as the wavelength of the light used, so that the whole of the diffraction grating is very compact indeed.

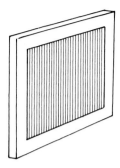

Figure 7.10   Diffraction grating

The formation of several spectra from one diffraction grating can be seen in the next section, and indeed the diffraction grating can profitably replace the prism mentioned in Section 7.17.

7.19   FORMATION OF MULTIPLE SPECTRA

Figure 7.11 shows how first order, second order and other orders of spectra are formed by rays of light whose paths differ by λ, 2λ, 3λ and so on. It should be appreciated that the wavelength of light is of the order of 6×10^{-7} metres whereas the distance AO might be of the order of 0.2 metres so that the diagrams are, of necessity, well out of scale.

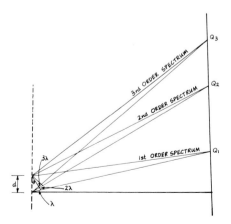

Figure 7.11   Formation of multiple spectra

7.20   DEMONSTRATION 6

To demonstrate a continuous spectrum.

Set up a slide projector containing Special Slide *C* (Section 1.3) and focus until a sharp image of the slit is seen on the screen. Darken the room as much as possible, the darker the better.

Take a diffraction grating of about 600 lines/mm and place it immediately in front of the projector lens, between the lens and screen.

A broad, wide continuous spectrum is seen on each side of the white central image of the slit. If the screen is wide enough a fainter secondary spectrum will be seen further out from the central image.

An even wider spectrum can be obtained if the diffraction grating is taped against the hypotenuse face of a right-angle prism. The prism should again be placed immediately in front of the projector lens.

Diffraction gratings are used extensively in obtaining spectra of the stars. They are more convenient to use than prisms and the spectra obtained can be more spread out.

7.21   PROJECT 42

To locate the first and second order spectra formed by a diffraction grating.

Data

The number of slits, or lines, per mm of the diffraction grating chosen will be assumed to be 500.

Wavelength of red light $= 7600 \text{ Å} = 0.76 \times 10^{-3} \text{ mm}$
Wavelength of violet light $= 3900 \text{ Å} = 0.39 \times 10^{-3} \text{ mm}$

For this project we shall assume that the distance between corresponding points on adjacent slits is equal to twice the width of one slit.

Calculate the width of one slit in millimeters. Draw vertically on the left hand side of a piece of paper a faint line. Rule in, near the top of one section, *AB* to represent one part of the portion of the diffraction grating which separates the slits. Making suitable allowance for the slit *BC* below *AB*, rule in part of the next portion *CD* which separates the slits. For all these a scale of $10 \text{ mm} = 0.2 \times 10^{-3} \text{ mm}$ will be suitable.

From the top point C of the bottom ruling draw a faint line AO horizontally for a reference. With centre C and radius that of the wavelength of red light, draw a semicircle on the right hand side of the diffraction grating using the linear scale given above. From point A, the corresponding point on the adjacent slit, draw the tangent to the semi-circle to touch the latter at point X. Join point C to point X and produce this line. Measure angle OCX. This is the direction of the red end of the first order spectrum. Compare this with 11°.

Repeat the above, but draw the semi-circle of radius equal to the wavelength of violet light to scale. Compare this with 5.6°.

So as not to confuse the drawing, repeat the above exercise on a second piece of paper, but to obtain the limits of the visible second order spectrum use radii for the semi-circles equal to twice the wavelengths of red light and violet light respectively. Compare the angles obtained with 22.3° for the red end and 11.2° for the violet end of the second order spectrum.

7.22 PROJECT 43

To construct a simple spectrometer, and to use it to observe a continuous spectrum and an emission spectrum.

We shall require a slide similar to Slide C in Figure 1.4 to form the spectrometer slit. We shall also require a diffraction grating of about 500 lines/mm preferably in bound slide form.

In whatever material is available (plywood is most suitable) make a box with a fixed bottom and a removable lid to the sizes shown in Figure 7.12. The shape of this box allows the light from the slit to pass through the diffraction grating at right angles; it also allows the first order spectrum to fall within the box and be approximately parallel to the scale.

Cut a rectangular hole and a circular hole in the box ends as shown in Figure 7.12. Cut a rectangular hole in the lid. Behind the rectangular hole inside the box fit the Slide C to form the slit. Behind the circular hole fit the diffraction grating inside the box with its parallel lines vertical and therefore parallel to the slit. On a piece of paper draw a scale a little longer than the hole in the lid, mark it 0 to 10 in equal divisions and stick it in position inside the box at the end beneath the hole in the lid. If possible put an eyepiece holder without lens over the circular hole outside the box to restrict the field of vision. Fit bottom and top making the box as light tight as possible.

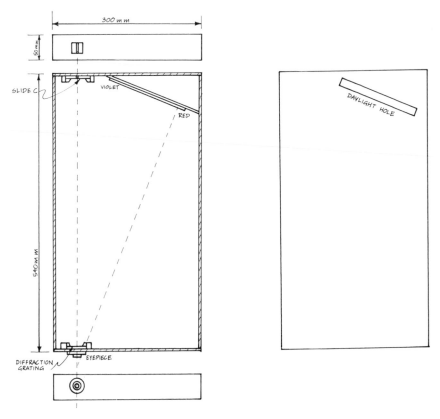

Figure 7.12   Simple spectrometer

Looking into the eyepiece, point the slit towards the daylight, sky or window. A continuous spectrum should be seen on the scale. Now point the slit at a fluorescent tube as used for domestic lighting, and observe a continuous spectrum but with bright lines crossing it at specific places. If the scale can be calibrated using monochromatic light, the elements causing the bright lines can be identified.

7.23   THE SPECTRUM OF A FLOURESCENT MERCURY VAPOUR LAMP

Using the simple spectrometer described in Project 43 to look at the spectrum of a fluorescent tube as used for domestic lighting, we observed a continuous spectrum crossed by a few emission lines. These lines are violet with an approximate wavelength of 436 nm, green with a wavelength of 546 nm and yellow (which, in an accurate instrument is a close pair) with a wavelength of 578 nm. The first two are quite bright and the yellow less so.

It is possible to observe more than one spectrum on each side of the undeviated light coming from the slit. The one nearest the slit is referred to as the first order spectrum, the next as the second order spectrum and so on.

The next project describes an easy way of determining the approximate wavelengths of each of the emission lines.

7.24 PROJECT 44

To measure the approximate wavelength of the emission lines in the spectrum of a mercury vapour fluorescent lamp.

The apparatus shown in Figure 7.13 consists of a low voltage (12 V) mercury vapour fluorescent lamp e.g. a camping light. This fluorescent

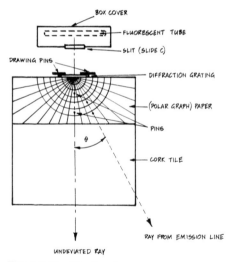

Figure 7.13 Wavelength measurement

lamp is covered by a box in the side of which is a narrow slit. The special Slide *C* shown in Figure 1.4 can easily be incorporated for this purpose if the box is made of cardboard. The narrow beam of light is directed on to a vertical diffraction grating which can be held in position by two drawing pins into the edge of a horizontal cork tile. A sheet of paper is laid on the tile, just touching the diffraction grating.

The spectrum will be observed from that side of the grating which is opposite to that of the light source. Looking at the spectrum, stick two pins, one behind the other so that they are both in line with the eye and the light coming from the slit. Also using two pins, line up the ray coming from the green line in the first order spectrum.

Draw the two lines indicated by the pairs of pins. Measure the angle θ between the two lines. Calculate the wavelength of the green light from the formula

$$\lambda = \frac{d.\sin\theta}{n}$$

where λ = the wavelength of the green light in nm   and n is the order of the spectrum, in this case 1.

Compare this with the value given in Section 7.23. The wavelengths of the other lines may be obtained similarly, and second order spectra may be used.

If polar graph paper is available the angle is more readily obtained. This exercise is best carried out in a dark room to avoid interference from other spectra caused by stray light.

7.25   THE HERTZSPRUNG—RUSSELL DIAGRAM

There is such a wide range of brightness and colour in the stars, that one might think that there could be no relation between them. If, however, we plot the absolute magnitude of the stars along the ordinate scale and the spectral type of the stars along the abscissa, we find that most of the stars fall along a fairly well-defined curve.

The stars which lie along this curve are referred to as main sequence stars. The stars at the lower right portion of the graph are stars in the early stages of development, not very bright and not very hot. As the star becomes older it also becomes hotter and brighter at first, and it will then have a position on the main sequence curve further up the curve to the left according to its mass.

There are some stars which do not lie on the main sequence curve. Those which lie above the curve are stars which have evolved further. They have left the main sequence and have travelled to the right in the diagram, becoming redder in the process. The radius of such a star also becomes larger, and this group of stars is a group of red giants. Since the stars do not all leave the main sequence, in their evolution, at the same point, these stars form a diffused area on the Hertzsprung-Russell diagram rather than a curve (Figure 7.14).

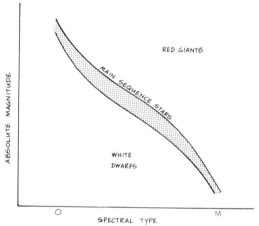

Figure 7.14 Hertzsprung-Russell diagram

Another group of stars lies in an area below the main sequence curve. These are the white dwarfs. They are stars in the last stages of evolution. After moving to the right from the main sequence stars, they move to the left becoming of variable brightness, cross the main sequence curve and finally end in the white dwarf area with very low brightness due to their small size. Because of this low luminosity relatively few white dwarfs have been detected.

7.26 PROJECT 45

(i)   To plot the Hertzsprung-Russell diagram.
(ii) To identify stars of the main sequence, red giants and white dwarfs.

Data

The Table below gives an extensive list of stars, their spectral class and absolute magnitude.

Set up a vertical axis for the absolute magnitude scale. Using a scale of 10 mm = 1 magnitude, mark this axis with a magnitude of 15 at the bottom and in intervals of 1 magnitude, up to a magnitude of − 7.

TABLE 7.6

	Star	Spectrum	Absolute magnitude
1	Naos	O5	−5.0
2	Alnitak	O9	−5.0
3	Mintaka	O9	−5.1
4	Cih	B0	−1.5
5	Alnilam	B0	−6.0
6	β Crucis	B0	−4.5
7	Acrux	B1	−2.7
8	ε Centauri	B1	−2.4
9	Murzim	B1	−3.6
10	β Centauri	B1	3.1
11	a Crucis	B1	−2.7
12	Spica	B2	−2.2
13	Bellatrix	B2	−2.9
14	Adhara	B2	−3.7
15	Achernar	B5	−2.6
16	Rigel	B8	−6.2
17	Regulus	B8	−0.7
18	Alpheratz	B8	−0.9
19	τ Corvi	B8	−0.5
20	Vega	A0	0.5
21	40 Eridani B	A0	11.1
22	Wolf 1346	A0	9.8
23	Hertzsprung	A0	11.0
24	δ Velorum	A0	0.1
25	Sirius A	A0	1.3
26	Bega	A0	0.6
27	Hyades No. 16	A0	11.2
28	Deneb	A2	−4.8
29	Fomalhaut	A3	2.1
30	Altair	A5	2.4
31	Ross 627	A5	13.6
32	Sirius B	A5	11.4
33	Altair	A5	−2.5
34	Canopus	F0	−4.6
35	Luyten 745-6	F0	14.3
36	Van Maanen's Star	F0	14.2
37	Scutulum	F0	−4.0
38	Caph	F2	1.7

	Star	Spectrum	Absolute magnitude
39	Procyon	F5	2.8
40	Sadir	F8	−3.5
41	α Centauri	G0	4.7
42	Capella	G0	−0.5
43	Sun	G1	4.8
44	Diphda	K1	0.8
45	Schedar	K0	−1.0
46	Arcturus	K0	0.0
47	40 Eridani	K0	6.0
48	Pollux	K0	1.0
49	σ Carinae	K0	−3.4
50	Eridani	K2	6.2
51	Almach	K3	−1.3
52	Fom	K3	−4.0
53	Aldebaran	K5	−0.5
54	λ Velorum	K5	−1.9
55	Groombridge 1618	K5	8.5
56	Kapteyn's Star	M0	11.2
57	Antares	M1	−2.4
58	Lacaille 8760	M1	8.6
59	Lacaille 9352	M2	9.4
60	Betelgeuse	M2	−5.6
61	Scheat	M2	−0.9
62	Lalande 25372	M2	10.2
63	CD−37° 15492	M3	10.3
64	CD−46° 11540	M4	11.3
65	Kruger 60	M4	11.9
66	BD+5° 1668	M4	12.2
67	Ross 614	M5	12.9
68	BD−12° 4525	M5	11.9
69	Barnard's Star	M5	13.2
70	CD−44° 11909	M5	12.8
71	Pegasi	M5	−1.5
72	Wolf 424	M6	14.3
73	Herculis	M8	−2.4

Set up a horizontal scale with its left hand end opposite magnitude 15 on the vertical scale. With spectral class O at the origin, mark off the other spectral classes B, A, F, G, K, M in that order to the right along this horizontal axis at intervals of 20 mm.

From the Table, plot each star on the graph, carefully marking each point with the number in the Table. These numbers have no astronomical significance, but they will assist in identifying the stars when the diagram is complete. The space between two adjacent classes should be divided into ten equal parts for the intermediate spectral classes.

When the chart is complete, a clear curve should be seen running from top left to bottom right of the diagram. A curved line can be drawn to include many of the stars, with some scatter on each side of the curve. The stars included by this curve and immediate neighbourhood are the main sequence stars.

A concentration of stars in the region around the point $(M, -1)$ will also be seen. These are the red giants, of which Arcturus and Aldebaran are two. Finally, there are a few stars together around the point $(A5, 12)$. These are the white dwarfs, and Sirius B is one of them.

7.27   THE RADII OF THE STARS

We have seen that stars evolve, and in doing so their physical size changes. If we assume that each star takes the form of a sphere radiating energy in the manner of a heated 'black body', it is possible to obtain an estimate of the radii of some of the stars, using the known laws of radiation and also the Hertzsprung-Russell diagram which we have just drawn.

According to Stefan's law, the radiation from a 'black body' is proportional to the fourth power of the temperature of the body. Temperature can be measured on various scales, but in this case the temperature used must be the absolute temperature. This is simply the temperature of the body in degrees Centigrade plus 273.

Also, for bodies of the same temperature, the total radiation will depend upon the surface area of the radiating bodies. For a sphere, the surface area is proportional to the square of the radius of the sphere. Thus, the radiation of stars is porportional to the squares of their radii.

The radiation from a star in a given time is its brightness. We therefore have a star X,

$$B_x \propto T_x^4.R_x^2$$

where $\quad T_x$ = absolute temperature of the star, and
$\qquad R_x$ = radius of the star

Thus, $\quad B_x = k.T_x^4.R_x^2 \quad$ where k is a constant of proportionality.

Similarly, for the Sun,

$$B_s = k.T_s^4.R_s^2$$

Dividing we have $\qquad \dfrac{B_x}{B_s} = \left(\dfrac{T_x}{T_s}\right)^4 . \left(\dfrac{R_x}{R_s}\right)^2$

But the ratio of the brightnesses is related in the equation for magnitudes

$$m_s - m_x = 2.5 \log_{10}\left(\frac{B_x}{B_s}\right)$$

where m_s and m_x are the absolute magnitudes of the Sun and star respectively.

Hence $\qquad m_s - m_x = 2.5 \log_{10}\left(\dfrac{T_x}{T_s}\right)^4 . \left(\dfrac{R_x}{R_s}\right)^2$

If we know the absolute magnitudes of the Sun and the star X, and can estimate the surface temperatures from the spectra, this equation will give us the ratio R_x/R_s, or the number of times the star is bigger than the Sun in radius.

7.28 PROJECT 46

(i) To plot on the Hertzsprung-Russell diagram, lines on which the ratio of the radius of the stars to the radius of the Sun is constant.
(ii) To determine approximately the radius of some stars.

Data

The Table below gives the surface temperatures of stars in the various spectral classes. Since the formula we have just deduced applies only to stars with temperatures up to about 7000 degrees absolute, we shall confine this Table to the spectral classes F, G, K and M.

TABLE 7.7

Spectral class	Temperature (K)
F0	7500
G0	5800
K0	4600
M0	3000

Temperature of Sun $= 5800\ K$
Absolute magnitude of Sun $= 4.8$

This project may be drawn directly on the Hertzsprung-Russell diagram constructed in Project 45 or more conveniently on tracing paper placed over the original diagram. In the latter case, trace the axes of the Hertzsprung-Russell diagram on to tracing paper together with the grid formed by the absolute magnitude and spectral class lines.

Taking all stars with a radius of 10 times that of the Sun, we have $(R_x/R_s)^2 = 100$.

For stars of spectral class $F0$ we have

$$(T_x/T_s)^4 = (7500/5800)^4 = 1.29^4 = 2.79$$

Thus a star with a radius 10 times that of the Sun and of spectral class $F0$ will have an absolute magnitude m_x given by

$$4.8 - m_x = 2.5\ \log_{10}(2.79 \times 100)$$
or
$$m_x = 4.8 - 2.5\ \log_{10} 279$$
$$m_x = 4.8 - 2.5 \times 2.446 = 4.8 - 6.12$$
or
$$m_x = -1.32$$

This point (absolute magnitude -1.32, spectral class $F0$) can now be plotted on the Hertzsprung-Russell diagram.

Using the *same* ratio of the radius of the star to the radius of the Sun, that is 10, and taking the temperature relevant to a $G0$ type star, repeat this calculation to find the corresponding absolute magnitude. Plot this on the diagram. Repeat for spectral classes $K0$ and $M0$. Draw a smooth curve through these points and mark it 10. Then all the stars on the Hertzsprung-Russell diagram which lie close to this line will have a radius approximately ten times that of the Sun.

We can, of course repeat this procedure for other ratios of the radius of the stars to the radius of the Sun, and so obtain a family of curves which are the loci of all stars which have a radius of so many times the radius of the Sun. If, however, we choose other multiples of 10, or sub-multiples of 10, so that the ratios taken for the radii are say 0.1, 1.0, 100, we note that the magnitude for each spectral class will change by 5 as we move from one ratio to the next. This can be seen from the re-arrangement of the formula, as below

$$m_x = m_s - 2.5 \log_{10} \left(\frac{T_x}{T_s}\right)^4 \cdot \left(\frac{R_x}{R_s}\right)^2 - 2.5 \log_{10} k^2$$

The last term is the change from one curve to the next along a constant spectral type line, since we have not adjusted the temperatures.

The last term is $-2.5 \log_{10} k^2$. If k is a factor of 10, then $\log_{10} k = \log_{10} 100 = 2$. Hence the last term in the equation for magnitude is $-2.5 \times 2 = -5$.

Thus if we increase the radius ratio by a factor of 10, we must add (-5) to the magnitude, and if we decrease the ratio by a factor of 10 we must add $(+5)$ to the magnitude.

To complete our diagram, take the point on the spectral line *F0* and radius ratio 10. Add -5 to the corresponding magnitude, and plot the result along the line *F0*. Repeat for the points on lines *G0*, *K0* and *M0*. Join the new points and label this curve 100.

Follow the same procedure for radius ratio 1.0, this time adding $+5$ to the magnitudes and labelling the curve 1.0. Taking the curve for radius ratio 1.0 as basis, add $+5$ to the magnitudes and so obtain the curve for radius ratio 0.1.

Lay the tracing paper over the Hertzsprung-Russell diagram so that the axes coincide. As a check on this method of determining the radius of the stars, we may compare the radius of each of the following stars with that given in other sources. A white dwarf, a main sequence star and a red giant have been chosen to give a range of values.

Sirius B has a radius of about 0.02 times that of the Sun,
Sirius A has a radius of 1.5 times that of the Sun,
Aldébaran has a radius of about 40 times that of the Sun.

In estimating for stars which lie between two curves, it should be remembered that the divisions will be wider nearer to the lower valued curve due to the logarithmic formula. The accuracy of this project depends on many factors, and is not, therefore, expected to be very great. The project does show, however, a method which is employed in determining the radii of some stars, with reasonable results.

7.29   THE DOPPLER EFFECT

The light which we receive from the stars and galaxies comes to us as a succession of waves.

Referring to Figure 7.15, the distance between adjacent crests is known as the wavelength (λ) of the light. The wavelength of any colour in the spectrum is different from that of any other colour, just as the wavelength of sounds varies according to the pitch of the notes. The number of complete waves which the light source creates in one second

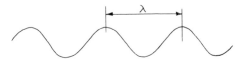

Figure 7.15   Wavelength

is known as the frequency (n) of the wave. Also the light energy transmitted by the waves has a velocity (c) of 3×10^5 km s^{-1}

In one second, n complete waves will have been generated of wavelength λ, so the leading wave will have travelled a distance $n \times \lambda$. But the distance travelled per second is (c) the velocity of the wave. Hence,

$$n \times \lambda = c \qquad [\text{i}]$$

We shall now consider the effect on the wavelength of light received by an observer O from a source S when there is relative motion between them.

In Figure 7.16, imagine that the source S is transmitting light waves of frequency n, and that these are being received by an observer O, both S and O being stationary.

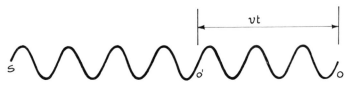

Figure 7.16   Doppler effect

Then, number of waves received in t seconds by observer O $= n.t.$
If we now add the compication that O is travelling
  towards S with a velocity v, then
    distance travelled by O in t seconds $= v.t.$

If the wavelength of the transmitted waves is λ, then

$$\text{number of waves in this distance} = \frac{v.t}{\lambda}$$

Thus it would appear that the moving observer O picks up a total number of waves in t seconds

$$= n.t + \frac{v.t.}{\lambda}$$

The observer judges the frequency of the source by the number of waves he receives in one second. Here he is clearly receiving a different number of waves per second than he would receive if he were stationary. Let the apparent frequency of the source be n', so that the number of waves received in time t second is $n'.t$.

Thus

$$n't = n.t + \frac{v.t}{\lambda}$$

or,

$$n' = n + \frac{v}{\lambda}$$

But $n = c/\lambda$ and $n' = c/\lambda'$ where λ' is the wavelength of the apparent signal received by O.

Hence

$$\frac{c}{\lambda'} = \frac{c}{\lambda} + \frac{v}{\lambda}$$

Re-arranging this equation, we have

$$\frac{\lambda}{\lambda'} = 1 + \frac{v}{c}$$

Subtracting unity from each side,

$$\frac{\lambda}{\lambda'} - 1 = \frac{v}{c}$$

or,

$$\frac{\lambda - \lambda'}{\lambda'} = \frac{v}{c} \qquad \text{[ii]}$$

Now for small velocities λ' will be almost the same as λ, and $\lambda - \lambda'$ is the reduction of wavelength as seen by the observer.

Thus,

$$\frac{\text{change in wavelength}}{\text{transmitted wavelength}} = \frac{\text{velocity of observer}}{\text{velocity of wave}}$$

A similar argument, with the same result, can be obtained for the case where the observer is stationary and the source moves *away* from the observer with velocity v. In this case there will be an *increase* in the wavelength as noted by the observer.

7.30   THE RED SHIFT

As we shall see shortly, the case of a stationary observer and a light source, such as a star or galaxy, moving away from the observer is by far of greater importance to the astronomer than any other case.

If the astronomer is receiving light which is producing either emission or absorption lines in a spectrum, then the wavelengths of these radiations will appear to be greater than the wavelengths if both source and observer were at rest. The lines will therefore appear to be moved towards the red end of the spectrum, that is, towards the greater wavelength end.

The displacement towards the red end of the spectrum is of great importance, and is known as the 'red shift'.

7.31   PROJECT 47

To determine the period of rotation on its axis of Jupiter using a spectrum of the sunlight reflected from this planet.

Figure 7.17 represents the spectrum of sunlight reflected from Jupiter when the planet was near opposition, that is, when the Earth and Jupiter were in line with the Sun but on opposite sides of it. The lines across Figure 7.17 are reproductions of some of the most prominent and clearly defined absorption lines on the actual spectrum. The vertical lines A and B represent the outermost edges of a diameter of Jupiter (see Figure 7.18). The marks on line C are calibration marks, the wavelengths of which are known accurately from laboratory tests.

Examination of the absorption lines will show that they are not at right angles to the spectrum but slope upwards from left to right. This is a consequence of the rotation of Jupiter. The lines on the left edge A are displaced towards the shorter wavelength end (blue) of the spectrum indicating that that edge of the planet is moving towards the Earth. Similarly the right edge B has a red shift indicating that that edge is receding from us.

We shall need to know the radius of Jupiter at the equator, which is 71 300 km.

With a fine pencil trace the centre band of the spectrum, and carefully mark in the lines. From the left hand edge of five of the most clearly defined lines draw a faint line across to the right so that it is at right

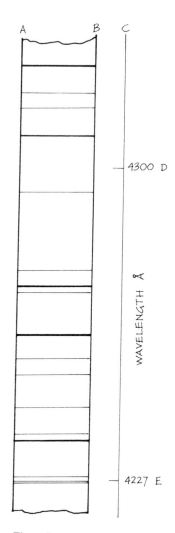

Figure 7.18   Jupiter references

Line illustration based on an original spectra
print © Lowell Observatory, Arizona,
U.S.A.
Reproduced by kind permission

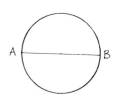

Figure 7.17   Jupiter spectrum

angles to the spectrum band. Carefully measure the total displacement of each line in millimetres, estimating to the nearest one-tenth of a millimetre. Add the five values obtained and divide the total by five thus arriving at an average value of the wavelength displacement in millimetres. To determine what each millimetre means in terms of wavelength we must use the calibration scale C.

Measure the distance DE in millimetres. By subtracting the marked wavelength at E from that marked at D and dividing by DE we obtain the scale of the spectrum in Angstrom units per millimetre.

Using this scale factor, determine the average wavelength shift in Angstrom units. We assume that the scale is uniform over the range taken since the range is small.

The spectrum is obtained from reflected sunlight and with the Sun and Earth in line with Jupiter this has the effect of making the spectral wavelength shift double what it should be. Also, the shift has been measured over the full width of the spectrum, corresponding to a diameter. We shall need the shift corresponding to a radius so that a factor of two is again involved. Further the axis of rotation of Jupiter is slightly tilted but we shall neglect the effect of this. Thus the actual wavelength shift over a radius is that measured divided by four.

We have seen in Section 7.29 that

$$\frac{\text{change in wavelength}}{\text{transmitted wavelength}} = \frac{\text{velocity of source}}{\text{velocity of wave}}$$

The transmitted wavelength can be taken as the average of that at D and that at E. The velocity of the wave is that of light, 3×10^5 km s^{-1}. Thus the tangential equatorial velocity (velocity of source) can be calculated.

Let the equatorial velocity obtained be v km s^{-1}. Then the angular velocity of Jupiter, $\omega = v / 71300$ radians s^{-1}. Since one revolution is equivalent to 2π radians, the time for one revolution will be $2\pi/\omega$ seconds or $2\pi/60\omega$ minutes.

Calculate this value, and compare it with the accepted value of the time for one revolution of Jupiter, namely 9 hours 50 minutes.

7.32 VELOCITY OF RECESSION

If the spectrum of a star or galaxy outside our own Galaxy is obtained with lines displaced towards the red end of the spectrum, it is possible, by comparing this spectrum with one obtained for bodies at rest, to obtain the velocity of recession of the star or nebula by using the formula [ii] (Section 7.29). It is found that distant stars and galaxies are receding from us at velocities proportional to their distance from us. This does not mean, however, that we are the centre of the universe. There is an analogy which demonstrates that all bodies can recede from each other in a universe that is expanding. It is that of currants in a sphere of dough. When the dough expands due to heat, the currants move in such a way that the distance between them increases, and any one currant observed from any other currant appears to be receding.

7.33   THE MOTION OF INDIVIDUAL STARS

The stars are so far away that any motion which a particular star may have is difficult to detect. This is because we rely on our being able to detect angular movement of the line joining us to the star. We have come to think of the stars as fixed, and indeed, we may regard them as such over relatively long periods of time. Nevertheless, the stars are moving relative to one another and relative to us, as is seen by the changing shapes of some of the constellations on star maps over the last two thousand years. Those stars which are relatively near to us will show up in different positions relative to the background of stars which are much further away, on examination of two photographic plates taken at the ends of a large interval of time.

In particular, a star with a large motion in the sky was discovered in 1916 by E. E. Barnard. This star appears to move a distance in the sky equal to the diameter of the Moon in about 180 years. Because of its large motion, photographs of Barnard's star have been taken continuously for a number of years. From these photographs it has become apparent that the star's motion is not in a straight line, but wanders from it in a curve which repeats itself about every 24 years. It has been deduced from this that Barnard's star must have a companion, and an estimate is that this companion is a planet similar in size to our planet Jupiter.

7.34   COMPONENTS OF THE MOTION OF A STAR

The motion we observe from photographs is really the projection of the true motion of the star on to the plane at right angles to our line of sight. This motion is called the proper motion or tangential component of the true motion. The other component of the true motion along the line of sight is referred to as the radial motion. These motions are illustrated in Figure 7.19.

If, as with Barnard's star, we are able to measure on photographic plates the distance moved by the star tangentially, we can calculate the tangential component of the true velocity of the star if we know the distance of

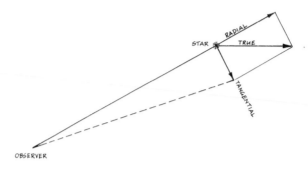

Figure 7.19   Component velocities of a star

the star and the scale of the photographic plate. We have seen earlier that the shift of lines in the spectrum of the star will give us the radial component of the true velocity. By compounding these vectorially, we are therefore able to arrive at a value for the true velocity of the star relative to us, and also the direction in space in which the star is moving.

7.35   PROJECT 48

(i)   To determine, from copies of two photographic plates, the proper motion of Barnard's star, and hence determine the tangential component of the true velocity of this star.

(ii) Given additionally the radial component of the true velocity of this star as obtained from its spectrum, to determine the true velocity in magnitude and direction.

(iii) Assuming that the velocity of this star relative to the observer remains the same, to determine the closest approach of Barnard's star to the Earth, and how long it will be before this event occurs.

Data

Copies of two photographic plates are provided (Figure 7.20). The one of 1924 shows the position of Barnard's star π and four reference stars marked 1, 2, 3 and 4. This plate has been exposed twice at different times, and so has two images of each of these stars.

The copy of the 1951 photograph shows the same stars. Each star has four images due also to multiple exposure of this plate at different times.

In both plates, the right hand image is the one corresponding to the specified date.

The top plate also has a line Ox inscribed on it. This gives the direction west to east. The other plate is orientated only approximately the same as the top plate.

The following constants will be required in subsequent calculations:

1 radian	$= 206\,265''$
1 year	$= 31.56 \times 10^6$ seconds
1 parsec	$= 30.9 \times 10^{12}$ km
Plate scale	$= 39.5$ seconds of arc per mm (as in Figure 7.20)

If graph tracing paper marked in millimetres is available, trace from Figure 7.20 the positions of the axes Ox and Oy, and mark on the tracing paper the right hand images of stars 1, 2, 3, 4 and π. If graph tracing paper is not available, ordinary tracing paper is at only a slight disadvantage. Now place the tracing over the lower copy of the photographic plate so that the right hand images of the reference stars 1, 2, 3, 4 coincide with those already on the tracing. Mark on the tracing the second image of Barnard's star.

Measure from the tracing, as precisely as possible, the distance in millimetres between the two positions of Barnard's star. This linear distance on the plate may be converted to the angle subtended at the observer by the plate scale. This, then, represents the proper motion of Barnard's star during the interval of 26.96 years between the two photographs. Calculate this proper motion, and compare it with the value of 10.27 seconds of arc per year.

With a protractor, measure the position angle of the proper motion. This is the angle between the direction towards north (in this case the top of the paper) and the direction in which the star is moving. This angle is taken as increasing in the direction towards the east (in this case, anti-clockwise on the paper).

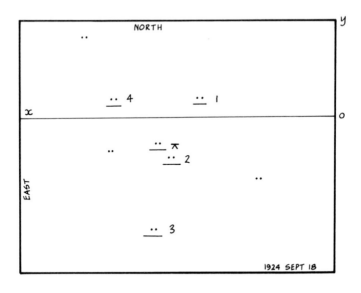

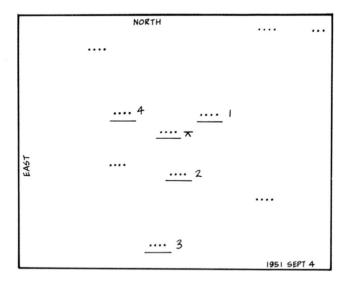

Figure 7.20   Proper motion of Barnard's star

Given that the parallax (see Section 7.7) of Barnard's star is 0.545″, we may now calculate the true velocity of the star relative to the observer, as given below:

Distance of Barnard's star = 1/(parallax)   = a parsec, say.
Distance of star = $a \times 30.9 \times 10^{12}$ km   = b km, say.
Proper motion as obtained above   = c'' per year

$$= \frac{c}{206\,265}\text{radians per year}$$
$$= d \text{ rad/year, say.}$$

Proper motion = $\dfrac{d}{31.6 \times 10^6}$ radians per second = e say.

Hence the distance moved tangentially per second
and this is the tangential velocity (V_t) required $= e \times b$ km,

The radial velocity (V_r) as obtained by the spectrum is -108 km/s, the negative sign indicating that this component velocity is towards us.

On a plain piece of paper, draw a line parallel and near to a longer edge to present the line of sight from the Earth to Barnard's star. Mark the position B of Barnard's star somewhere on this line, near to one of the shorter edges. Mark also on this line the position of the Earth E, using a scale of 100 mm = 1 parsec, at a distance of a parsecs from B. From B measure the radial velocity BR towards the Earth E, using a scale of 1 mm = 1 km/s. From B draw a second line BT to represent the tangential velocity at right angles to BE, and using the same scale. Complete the rectangle $TBRA$.

Then the line BA represents the magnitude, to scale, and the direction of the velocity of Barnard's star relative to the observer. Measure this diagonal and record the true velocity V.

Next, draw a line from E at right angles to BA to intersect BA in N. Then N is the point at which Barnard's star will be closest to the Earth. Record this minimum distance, and consider whether or not the star will be closer to the Earth when at N than the present star Alpha Centauri, which is 1.33 parsecs from the Earth.

Travelling at its present velocity, how long will it take Barnard's star to travel 1 parsec? Compare this with 6.89×10^4 years. Calculate also how long it will take for Barnard's star to reach its closest approach to the Earth.

7.36   BINARY STAR ORBITS

We have previously mentioned the binary stars, that is, a system of two stars revolving about their common centre of mass. Relative to one of the stars, the other star of the system moves in an elliptical orbit like a planet orbiting the Sun.

The binary stars are important in that they allow us to determine the total mass of the two stars by consideration of their orbit, the period of the orbit and the parallax of the stars in the system.

The movement of one component B of a binary system around the other component A which we see, is not the true shape of the orbit unless the orbit is perpendicular to the line of sight. The orbit which we observe is known as the apparent orbit.

If the star system is well removed from other stars, then the motion is that of two bodies under their own gravitational forces, and the true orbit of B relative to A will be an ellipse with A at one focus. The calculation of the sum of the masses of A and B can be carried out only if we can deduce the semi-major axis of the true orbit from the apparent orbit. Project 49 shows how this can be done graphically. The proof of the construction, which is known as Zwier's method, is rather involved and will not be given here. Readers with a good knowledge of the geometrical properties of the ellipse may be interested to read the basis of the method in W. M. Smart's *Spherical Astronomy*, published by the Cambridge University Press.

The period of time necessary for a star of a binary system to revolve round its companion, can be obtained by observation, but as the period is often a long one, the observations may have to be carried out over several lifetimes.

Another reading which has to be taken at intervals, is the angular separation of the two components. This is really the angle which the two stars subtend at the observer, and is measured in seconds of arc. If the linear distance corresponding to this is required, we must multiply the angular separation, converted to radians, by the distance of the binary system from the observer.

Finally, we need to know the position of component B relative to component A. This is specified as the position angle, the angle between the line pointing to north and the line joining A to B. The direction of measurement of the angle is from the north towards the east.

7.37 PROJECT 49

(i) To plot the apparent orbit of one component of a binary star system relative to the other component.
(ii) To deduce the semi-major axis of the true orbit of the system, and the angle which the plane of the true orbit makes with the plane perpendicular to the line of sight.

Data

The Table below gives the position angle and angular separation of the components of the binary star Zeta Herculis for one revolution.

TABLE 7.8

Date	Position angle (°)	Separation (″)
1931.9	352.0	0.58
1935.9	227.0	0.92
1937.6	207.6	1.07
1939.8	188.6	1.14
1941.8	170.3	1.16
1943.8	154.3	1.20
1945.8	138.7	1.26
1947.6	125.8	1.32
1949.7	115.2	1.39
1949.8	(117.3	1.41)
1951.7	101.5	1.48
1953.7	91.9	1.55
1955.8	82.0	1.59
1957.6	73.8	1.58
1959.6	64.2	1.52
1961.7	52.6	1.36
1963.7	37.7	1.09
1965.7	7.2	0.70
1967.4	305.5	0.49
1968.2	270.5	0.57
1969.0	248.4	0.73
1969.3	242.7	0.78
1969.8	233.6	0.87
1970.7	222.1	0.97

Set up perpendicular axes intersecting in S, approximately in the centre of the paper. S represents the primary component of the binary system. Assume that the north direction is along the axis pointing vertically downwards. From this reference line, draw radial lines from S at angles corresponding to the position angles, measuring them in an anti-clockwise direction. On each position line, measure from S a distance proportional to the separation of the two components, using a scale 40 mm = 1 second of arc. Label each radius with the correct date.

Join the points so obtained with a smooth curve, which will be the apparent orbit. Locate the centre of the elliptical orbit by drawing two parallel chords. Join the mid points of these chords, and produce this line in both directions to meet the ellipse in G and G_1. The centre of the ellipse C bisects GG_1.

Join CS and produce it to cut the ellipse in A_1. Produce SC to cut the ellipse in D_1. Then the eccentricity of the true orbit is given by

$$e = \frac{CS}{CA_1}$$

Measure CS and CA_1 and obtain this ratio. Compare the value obtained with 0.48.

Calculate $$k = \frac{1}{(1-e^2)^{\frac{1}{2}}}$$

Draw a chord UW parallel to A_1D_1. Bisect this chord in V. Join V to C and produce in both directions to cut the apparent ellipse in B_1 and E_1. Draw a series of chords parallel to B_1E_1 across the ellipse (about six, equally spaced, will be enough) to cut the ellipse in typical points T and T_1 and the chord A_1D_1 in R. Extend RT to X such that $RT/RX = k$ calculated above. Similarly extend RT_1 to X_1. Repeat for the other chords drawn last. Join up the typical points X and X_1 with a smooth curve. This will be an ellipse, but it is not the true orbit. However, its major axis has the same length as that of the true orbit.

Locate the two points A_2 and D_2 which are furthest from C, by trial and error. Join A_2CD_2. This is the major axis of the second ellipse, and also of the true orbit. Measure CA_2 and compare with 1.37".

Finally, draw CB_2 at right angles to CA_2 to cut the second ellipse in B_2. Then the angle i which the plane of the true orbit makes with the plane perpendicular to the line of sight is given by

$$\cos i = \frac{CB_2}{CA_2}$$

Measure CB_2 and CA_2 and thus obtain $\cos i$.

Using tables or a calculator, obtain the value of i and compare with 50°.

7.38 ECLIPSING BINARY STARS

The light energy received from some stars varies with time. These stars are referred to as variable stars. Some of these vary in brightness because the light output changes due to some phenomenon such as pulsation. Others, however, vary in brightness simply because they are real binary systems with the plane of their orbit nearly along the line of sight. As one star revolves around the other, we get eclipses taking place. With one star in front of the other, some of the light is cut out, thus diminishing the observed brightness. When the stars appear to be separated, we see a brighter object which has the brightness of the sum of the individual components approximately.

Because the light variation is due to eclipsing, we would expect a periodic variation of brightness with time. Thus, even though the components may be so close together that they cannot be separated visually, we can tell from the light curve that the system is binary.

An inspection of the light curve will reveal more than this. The shape of the curve will tell us whether the system has partial eclipses or whether they are total or annular. We may also deduce the relative brightnesses of the two components. Coupled with spectroscopic measurements, the size of the orbit and the mass of each component can be obtained, as we have seen.

Project 50 takes an eclipsing binary system, and for different relative brightnesses of the two components, shows how the approximate light curve can be obtained. In a similar way, the light curve for a totally eclipsing system is constructed, and the differences between these light curves can then be examined.

7.39 PROJECT 50

(i) To plot typical brightness curves for different eclipsing binary star systems.
(ii) To note the difference in the curves for a totally annular eclipsing system and a partially eclipsing system.

(iii)   To note the effect on the light curve of the relative brightnesses of the component stars of the system.

Draw a circle to represent the larger star. This may conveniently be 25 mm radius. The method to be described depends on area measurement, and unless a planimeter is available, it will be found profitable to carry out the drawing on squared paper. Draw a diameter of this circle, and a line parallel to it but 20 mm from it, to represent the path of the centre of the smaller star.

With centre at centre of the large circle, and radius 37.5 mm, cut the second straight line with an arc. This will give the centre of the smaller star when it is apparently just touching the larger star if the radius of the smaller star is taken as 12.5 mm. Draw this circle.

From the centre of the larger star, drop the perpendicular on to the path of the smaller star to cut it in N. Measure the distance from the centre of the smaller star to N.

We shall assume that, for simplification, the second star is moving uniformly across the first star, so that this distance will represent time units. Measure the area of both circles and add them.

Next, let the centre of the smaller star move to 25 mm from N. Measure the total area of star showing. Repeat for distances of 20, 15, 10, 5 and 0 mm from N. Plot time units horizontally and total area, which is proportional to total brightness, vertically. Suitable scales are 10 mm = 20 time units and 10 mm = 5 units of area (brightness).

Since we are assuming in the first case that the brightness per unit area is the same for each star, the light curve will be repeated exactly for the passage of the smaller star behind the large star. In between, the light curve will be constant at the sum of the areas (brightnesses).

In the second case we shall assume that the brightness of the small star is twice that of the large star. The project can be carried out exactly as before, but in calculating the total brightness at any time we must add twice the area of the small star which is showing to the area of the large star which is showing.

The above two parts of the project will give light curves for cases where there is a partial eclipse of one of the stars. In the next part of the project we draw the path of the small star's centre only 7 mm from the diameter of the large star. In this way we shall simulate a total eclipse of the small star. The whole of the instructions given previously can be carried out, so that we obtain first the light curve when the brightness per unit area for each star is the same, and then the light curve when the brightness of the small star per unit area is twice that of the large star.

Comparisons can now be made between the light curves of the different cases, and we can see how, given such light curves for a variable star, we could deduce whether or not the star is, in fact, an eclipsing binary system and, if so, whether it is partially or totally eclipsing and what are the approximate relative brightnesses of the two components.

7.40   THE MASSES OF THE STARS

In 1924 Eddington discovered a theoretical relationship between the masses of the stars and their absolute magnitudes.

If stellar mass is plotted on a logarithmic scale against absolute magnitude, a graph, which is sensibly straight over most of the range, is obtained. The greater the mass of the star, the greater is its absolute magnitude and therefore also its brightness.

We can obtain practical data from some of the binary stars to test this relationship. The sum of the masses of the stars in a binary system can be obtained by applying the exact form of Kepler's third law, namely

$$G(M+m) = 4\pi^2 . \frac{a^3}{T^2}$$

where M, m = masses of the primary and secondary components
$\quad\quad a$ = semi-major axis of the true orbit
$\quad\quad T$ = period of revolution
$\quad\quad G$ = constant of gravitation

We verified this law in Project 4 for the special case of a planet revolving round the Sun, when the mass of the planet could be neglected compared with that of the Sun. In the case of binary stars, the two components usually have comparable masses, and so the full equation must be used.

Thus, for a binary system

$$G(M+m) = 4\pi^2 \frac{A^3}{T^2}$$

and for the Earth revolving found the Sun

$$G.m_s = 4\pi^2 \frac{a^3}{1^2}$$

where

m_s = mass of the Sun

a = radius of the Earth's orbit round the Sun, that is, its semi-major axis

and the time of revolution is unity, that is, 1 year.

By division, we have

$$\frac{M+m}{m_s} = \left(\frac{A}{a}\right)^3 \frac{1}{T^2}$$

Now, if d is the distance of the binary star from the Sun, and a is the angle subtended at the Sun by the semi-major axis of the true binary orbit, then

$$A = d.a$$

Similarly, if p is the parallax of the binary system,

$$a = d.p$$

By division again

$$\frac{A}{a} = \frac{a}{p}$$

Thus

$$\frac{M+m}{m_s} = \left(\frac{a}{p}\right)^3 \cdot \frac{1}{T^2}$$

Since a, p and T can be measured in many cases, the sum of the masses of the binary components can be determined in terms of the mass of the Sun.

In some cases, the ratio of M to m can be obtained either through spectroscopic observation or through the deviation of the motion of one of the binary components from the motion expected from a single star. Thus, with the ratio and the sum of the masses, each individual mass can be obtained.

In particular, for use in Project 51, Sirius A has a mass of 2.44 times that of the Sun, and Sirius B has a mass of 0.96 times that of the Sun. Kruger 60A has a mass of 0.24 times that of the Sun and Kruger 60B has a mass of 0.16 times that of the Sun.

7.41 PROJECT 51

(i) To plot a graph relating the absolute magnitude of stars and their masses.

(ii) To determine from the graph the masses of a few stars.

Data

The mass of Sirius A is 2.44 times that of the Sun.
The absolute magnitude of Sirius A is 1.3.

Set up rectangular axes such that the absolute magnitude lies along the abscissa, ranging from -4 to $+16$ to a scale of 10 mm = 1 magnitude, such that the ordinate represents the logarithm to base 10 of the mass, and such that 50 mm represents 1.0 units of $\log_{10} M$, where M is the mass of any star in terms of the mass of the Sun.

The mass of the Sun is 1 times its own mass. The value of M is therefore 1, and $\log_{10} 1 = 0$. The Sun's absolute magnitude is 4.8. From this information, mark the position of the Sun on the graph.

The mass of Sirius A is 2.44 times that of the Sun. Its absolute magnitude is 1.3. Plot the position of Sirius A on the graph after finding the value of $\log_{10} 2.44$.

Draw a straight line through the two points on the graph. This is the mass-luminosity relationship.

Given that the absolute magnitude of Kruger 60A is 10.1, and that Capella A has absolute magnitude of -0.5, find from the graph the value of $\log_{10} M$ for each of these stars, and by taking the antilogarithm of each value find the mass of each of these stars compared with the mass of the Sun. Compare the masses obtained with 0.24 for Kruger 60A and 4.2 for Capella A.

8

Galaxies and nebulae

8.1 THE GALAXY

Our Sun is one star of a very large number of stars (more than 100 000 million stars) which are rotating together in the manner of a spinning disc. The disc has a more or less spherical bulge at its centre. Since the Sun is located some way from the centre, the number of stars of this group which we see from our solar system depends upon the direction in which we look. In particular, if we look towards the centre, we see projected against the celestial sphere a band of many stars, which we call the Milky Way because the impression to the naked eye is one of a more or less diffuse band of light.

The group of stars with which we are associated is known as the Galaxy, written with a capital G to distinguish it from the millions of other galaxies. The best known of the latter is the Andromeda galaxy, which can be seen on a clear moonless night with the naked eye. Our own Galaxy is similar in structure to the Andromeda galaxy, photographs of which show that it is of a rotating spiral form.

8.2 ASTRONOMICAL DISTANCES

The determination of the distances of planets, stars and galaxies has, of necessity, to be mostly by indirect methods. The methods used for the nearer bodies fail when applied to the more distant bodies, but the distances which have been determined for the nearer bodies can be used as stepping stones for bodies further away.

The distance of the Moon from the Earth can now be determined very accurately by radar and by the laser beam equipment left on the Moon by the Apollo crews. The distances from the Earth to the inner planets,

Mercury and Venus, and of Mars, can also be obtained by radar. Indeed, some of the contours on Venus, which are continually obscured by a carbon dioxide atmosphere, can be mapped by radar as can surface features on the Moon and other planets mentioned. Radar consists essentially of sending out pulses of electrical energy and collecting the reflected rays after they have struck the target. The interval of time between the sending out of the signal and the reception of the reflection can be measured very accurately. This time multiplied by the speed at which the signal travels, which is the speed of light, will give the total distance travelled. This is twice the distance being measured, although the true situation is more complicated than that described.

We have seen that by using Kepler's third law we can determine the semi-major axes of the orbits of all the planets provided we know the semi-major axis and period of revolution round the Sun for one of them and can observe the periods of the others. Thus the size of the solar system is determined. In particular the semi-major axis of the Earth's orbit is known accurately.

To reach the nearest stars we can use the method of parallax introduced when we were discussing stellar magnitudes. The basis of the method is the same as for triangulation surveys on Earth. A baseline AB of known length is taken and the angles which the baseline makes with each of the lines joining A to the target and B to the target C are measured (see Figure 8.1). A simple application of trigonometry will give the distance

Figure 8.1   Triangulation

of the target from A and from B. The nearest stars are, however, so far away that no Earth-bound baseline is long enough to show a measurable difference in the directions AC and BC. Consequently, observations are made when the Earth is at one end of the major axis of its orbit and again six months later. The baseline being used here is the major axis of the Earth's orbit. It will be appreciated that with such large distances (the nearest star has a parallax on only 0.75″, making it 1.3 parsecs or 4.4 light years away) exact values cannot be expected and as the distances are increased so is the uncertainty of measurement.

The method of parallax will take us out with some uncertainty to a distance of about 100 parsecs, beyond which the parallaxes become too small to measure accurately.

The next stepping stone relates to stars which are moving as a group, the so called moving clusters, of which the Hyades is one. Although stars appear to be stationary they are moving, and the motion at right angles to the line of sight can be measured. Couple this with the near certain assumption that the stars in the group are at the same distance from the observer and this provides us with a method of determining the distance of the cluster (see Project 52).

To extend the distance measurement still further we need to be able to observe individual stars with known characteristics. The *RR* Lyrae stars are variable in such a way that they have a recognisable brightness-time curve. By observing *RR* Lyrae stars at distances which we can determine we can, from their mean apparent magnitude and distance, calculate their mean absolute magnitude (see Section 7.13). It is found that the absolute magnitudes of *RR* Lyrae stars are about the same. It follows that if we observe *RR* Lyrae stars further away, we can measure their mean apparent magnitudes, we can assume their mean absolute magnitudes and hence estimate their distance.

Unfortunately the absolute magnitude of *RR* Lyrae stars is only about 0.5 and these stars can only be observed within our own and some of the nearer galaxies. We can, however, extend this method using Cepheids. These variable stars again have a recognisable brightness-time curve and there is a definite relation between the length of the fluctuation period and the mean apparent magnitude. Again, if we can observe Cepheid stars which are at a known distance we can determine their mean absolute magnitude and so relate this to the period of fluctuation. In reverse the mean apparent magnitude and period will lead to mean absolute magnitude and thence to distance. Cepheids are more useful than *RR* Lyrae stars because they are brighter and therefore can be seen at greater distances. A note of caution is needed here in that there are two types of Cepheids, of Population I and of Population II. The former are about 1.4 magnitude brighter than the latter. Such is the nature of distance determination in astronomy that when the existence of two Populations of Cepheids was recognised it led to a revision of some distances by a factor of two. Project 53 uses the Cepheid method of distance determination, which takes us out to a distance of about 1.5 million parsecs.

The appearance of novae gives us another way of determining distance since the brightness of novae makes these objects readily visible. The method is described in Project 54.

To extend our probing still further we may use similar methods but substitute the brightness of whole galaxies, in which single stars cannot be resolved, for the brightness of individual stars.

Finally we come to the well known red shift of distant galaxies, the use of which will extend our reach into the universe about 5000 million light years. More about this will be said later (Section 8.15 and Project 57).

Perhaps the telescope now being constructed for use in satellites above the Earth's atmosphere will take us very near to the edge of our observable universe.

8.3   STAR CLUSTERS

Over the area of the sky we find, in some cases, many stars grouped together so closely that the overall effect is that of a ball of light. These are the globular clusters. They are spread throughout the volume of space which is roughly the sphere of which the centre is that of our Galaxy, and whose radius is also that of our Galaxy.

In other cases there are groups of stars, only relatively few in number, which are connected dynamically. These open clusters, as they are known, occupy a limited volume of space, and the individual stars are usually to be found fairly close together in the sky. These stars are characterised by their common motions; it appears that the members of such a cluster move with identical velocities relative to the Sun. In particular, they travel in parallel paths in space.

Only the nearest of the open clusters have had their velocities measured. Such a cluster is the Hyades, a group of stars in the constellation of Taurus. The motions of these stars show up significantly over a relatively short period of time.

Using the same reasoning as that for meteor streams (Section 4.12) we deduce that, since the stars in a dynamically associated cluster are actually moving along parallel paths, we observe them as moving towards a common point in the sky, which in this case is called the convergent. The effect is much less obvious than in the case of meteors because it is infinitely slower. A meteor may cover several degrees of the sky in about one second, but stars in an open cluster cover something like one tenth of a second of arc in a year.

Before passing on to the next project, which allows us to determine the distance of stars in an open cluster, we need to consider one or two points connected with the celestial sphere.

8.4   ANGLES AND DISTANCES ON THE CELESTIAL SPHERE

In Figure 8.2, AB represents an arc of the equator of the sphere subtending an angle a at the centre of the sphere. If the radius of the sphere is r, then the arc $AB = ra$. It is customary, since it simplifies astronomical calculations, to think of the radius of the sphere as unity, so that we can now refer to the arc AB as a. This will apply to arcs of other circles of the sphere whose planes pass through the centre of the sphere. Circles such as these are called great circles. Thus the arc BC may be referred to as δ.

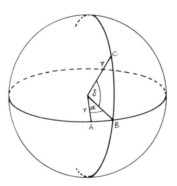

Figure 8.2   Celestial sphere distances

Looking now at Figure 8.3, where the plane of the smaller circle is parallel to the plane of the larger great circle, it is clear that in the triangle DEF,

$$r_1 = r \cos \delta$$

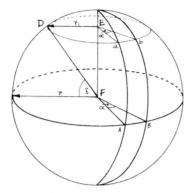

Figure 8.3   Distances at declination δ

The circumference of the smaller circle is therefore $2\pi r_1 = 2\pi r \cos \delta$, while that of the larger circle is $2\pi r$. It will also be seen that the arcs ab and AB which subtend equal angles a at the centres of their respective circles, are connected by the relation

$$ab = AB \cos \delta$$

or if we take the radius of the larger circle as unity

$$ab = a \cos \delta$$

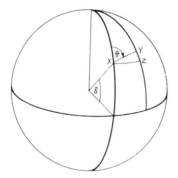

Figure 8.4   Displacement components

Applying this to our special case of a star X of an open cluster (Figure 8.4) which moves to Y in a certain time through a distance μ seconds of arc, we see that if μ_a is the resulting change in right ascension, then

$$XZ = \mu_a \cos \delta$$

Also, if μ_δ is the resulting change in declination, then

$$ZY = \mu_\delta$$

Again referring to Figure 8.4, the angle θ is called the position angle of the proper motion XY. It is measured relative to the great circle passing through X, and is in the direction shown.

Since we are dealing with an extremely small part of the sky, XYZ may be considered as a plane triangle with a right angle at Z, and angle $YXZ = 90° - \theta$.

Hence $\mu_a \cos \delta = HZ = \mu \sin \theta$ [i]

and $\mu_\delta = YZ = \mu \cos \theta$ [ii]

Dividing [i] by [ii],   [iii]

$$\tan \theta = \frac{\mu_a \cos \delta}{\mu_\delta}$$

8.5   VELOCITY COMPONENTS

In Figure 8.5, the point X represents one of the stars in the open cluster, and the point C represents the position of the convergent of the cluster. S represents the Sun. The angle β is the angle subtended at S by the arc XC of a great circle, and if we follow our convention of regarding the radius of the sphere as unity, then the distance XC can be said to be β.

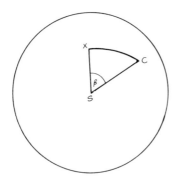

Figure 8.5   The Sun, star and convergent

In Figure 8.6 the velocity V of star X is parallel to the line SC, the line joining the Sun to the convergent (see Figure 4.13, Section 4.12).

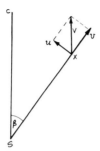

Figure 8.6   Open cluster velocity components

Now V can be resolved into two components: the transverse component $u = V \sin \beta$ and $v = V \cos \beta$ which is known as the radial component, both of which lie in the plane CSX. The velocity v can be determined from the shift of lines in the spectrum.

We are now in a position to start Project 52.

8.6   PROJECT 52

(i) To obtain the mean velocity of stars in an open cluster, in this case the Hyades.

(ii) To determine the distance of stars in the Hyades.

Data

The Table below gives the right ascension and declination of stars in the Hyades. The radial component of the stars' velocities are given, and the quantities $\mu_a \cos \delta$ and μ_δ have also been given. The column of magnitudes shows that all of these stars are visible to the naked eye.

First convert the values of right ascension in hours and minutes to degrees and minutes of arc. For this conversion we must remember that 24 hours is equivalent to 360°, so that 1 hour $\equiv 15°$ and 1 minute $\equiv 15'$.

Set up axes on a sheet of graph paper, with the right ascension as abscissa and declination as ordinate. For both scales make 10 mm $\equiv 2°$, and start the right ascension from 52° at the extreme left with about 96° at the right hand end of this axis. Start the declination at 6° at the bottom, rising to about 24° at the top. Using the data table plot on the graph sheet the position of each star.

TABLE 8.1

Star	R.A.	Dec.	Mag.
1201	3h 47m	+17°02′	5.97
1279	4 02	+14 54	6.01
1283	4 03	+19 21	5.49
1319	4 10	+15 09	6.32
1373	4 17	+17 08	3.76
1389	4 20	+17 42	4.30
1392	4 21	+22 35	4.29
1394	4 21	+15 23	4.48
1409	4 23	+18 58	3.54
1411	4 23	+15 44	3.85
1427	4 25	+15 59	4.78
1473	4 33	+12 19	4.27
1620	4 57	+21 27	4.64

Star	$\mu_a \cos \delta$	μ_δ	v(km/s)
1201	+0.141″	−0.028	+35
1279	+0.131	−0.024	+36
1283	+0.103	−0.032	+24
1319	+0.115	−0.029	+37
1373	+0.105	−0.031	+38
1389	+0.107	−0.029	+35
1392	+0.100	−0.047	+35
1394	+0.111	−0.023	+41
1409	+0.106	−0.038	+39
1411	+0.101	−0.028	+40
1427	+0.105	−0.028	+38
1473	+0.099	−0.012	+45
1620	+0.056	−0.043	+42

Next, draw up a Table with following headings:

Star	$\tan \theta$	$\theta°$	β	$\sec \beta$	v (km/s)	V (km/s)

Calculate $\tan \theta$ from $\tan \theta = \mu_a \cos \delta / \mu_\delta$ for each star, and using a good set of tables such as *Chambers' Six Figure Mathematical Tables,*

Volume 2, Natural Values, determine the value of θ for each star. Care must be taken to decide the trigonometric quadrant for θ. In this case, sin θ is positive since all the values of μ_a cos δ given in the Table are positive. Also the value of cos θ is negative since all the values of μ_δ are negative. Thus θ lies in the second trigonometric quadrant, and is therefore between 90° and 180°.

Using these position angles, and a protractor and ruler, draw on the graph the apparent direction of the proper motion of each star. For this exercise, the data presented does not have a high accuracy, and to a first approximation the curvature of the celestial sphere can be neglected. Extend these lines of direction to cut the other lines and mark with a clear point every intersection. There will be some scattering of these points of intersection, so we determine the point of convergence of the cluster by drawing an oval to include that part where the points of intersection have their highest concentration. Take the position of the centre of this oval as being the point of convergence C. Read off the right ascension and declination of this point.

Compare these values with right ascension = 6 h 10 m
declination = +7°

We have seen that, if β is the angular distance of any star X from the point of convergence C, then the radial velocity v is given by $v = V \cos \beta$, where V is the velocity of the star, this being common to all stars of the cluster. Hence $V = v \sec \beta$.

For each of the plotted stars, measure the angle β. Since we have made the scales of both axes of the graph the same, β is merely the linear distance from the star X to the convergent point C. Again using Tables, find and tabulate the corresponding values of sec β, and using $V = v \sec \beta$ determine the value of V for each star. Add all the values of V, and divide by the number of stars involved, and so obtain a mean velocity of the cluster.

Compare this with the value 42 km/s

If a star is at a distance of d km and its annual proper motion is μ, then it follows from arc = radius × subtended angle and because μ is very small, that

$$\mu = \frac{u \times 3.156 \times 10^7}{d}$$

where u is the transverse velocity of the star in km/s and 3.156×10^7 is the number of seconds in a year.

Hence
$$d = 3.156 \times 10^7 . \frac{u}{\mu}$$

Now the parallax p of a star is given by

$$p = \frac{a}{d}$$

where a is the radius of the Earth's orbit, namely 1.497×10^8 km.

Hence,
$$p = \frac{1.497 \times 10^8}{3.156 \times 10^7} \frac{u}{\mu} \qquad \qquad [iv]$$

p and μ are normally expressed in seconds of arc.

Now
$$u = V \sin \beta \qquad \qquad [v]$$

and
$$\mu = \sqrt{(\mu_a \cos \delta)^2 + \mu_\delta^2} \qquad \qquad [vi]$$

Hence, by knowing V, already found, and β, already measured, and by calculating u and μ from [v] and [vi] above, the parallax of any star in the open cluster can be determined.

Using this information, find the parallax, and hence the distance, of star 1373 in light years.

Compare this with a general figure for the distance of the Hyades of 120 light years.

8.7   PROJECT 53

(i)   To plot the absolute magnitude of δ Cepheid stars in the Small Magellanic Cloud against the period of light variation assuming a distance for this cloud.

(ii) To deduce, from this, the distance of the variable star δ Cephei from which these stars take their name.

Data

The apparent magnitudes of a few Cepheid type stars and their corresponding periods of light variation are given below, all stars being in the Small Magellanic Cloud whose distance from the solar system is 200 000 light years.   1 parsec = 3.26 light years.

Using the formula (or nomograph in Section 7.15) $M = m + 5 + 5 \log_{10} p$ for each of the stars in the Table below determine the corresponding absolute magnitudes M. Their parallaxes p, in seconds of arc, may be

TABLE 8.2

Apparent magnitude (average)	Period (days)
14.95	10.0
14.2	22.4
13.8	31.6
13.0	79.4

obtained from the assumption that since they are all in the Small Magellanic Cloud they are all at the same distance.

Draw a graph of absolute magnitude M as ordinate, to a scale of 10 mm = 1 magnitude with a range of magnitude -7 at the top and zero at the bottom, against $\log_{10} P$ as abscissa, where P is the period in days, to a scale of 10 mm = 0.2 $\log_{10}$ units, ranging from -0.2 at the left hand end to 2.2 at the right end.

Determine from Figure 8.7 the period P_δ of the star δ Cephei. Obtain $\log_{10} P_\delta$, and from the graph read the corresponding absolute magnitude

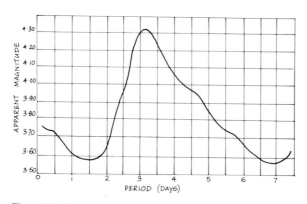

Figure 8.7   Variations of apparent magnitude of δ Cephei

M_δ. Calculate m_δ the "average" apparent magnitude of δ Cephei from the maximum and minimum values in Figure 8.7. Using the formula $M_\delta = m_\delta + 5 + 5 \log_{10} p_\delta$ determine the parallax p_δ of δ Cephei and hence the distance of this star in light years. Compare this with 650 light years.

8.8   NOVAE

Among the stars whose brightness is not constant are a few stars, which are often not very bright in normal circumstances, but which suddenly increase in brightness rapidly in the time of a few days. At that time these stars may be among the brightest in the sky. After this period, the brightness decreases, fairly quickly at first, and then at a rather slower rate which is roughly constant. These stars are called novae, or new stars. New stars here simply means that they are much more easily observed, and could well give the impression of appearing where no star had been before.

Some of the stars in this category are sufficiently close to us to allow us to determine their distance by the parallax method (see Section 8.2). Since we can observe the apparent magnitude of these stars when they are at their maximum brightness, we can, using the known distance, determine their absolute magnitudes at maximum brightness. It is found that most of these absolute magnitudes lie between -5 and -10, with a preponderance of between -6 and -7. Thus if we observe a number of novae in another galaxy, the chances are that they have absolute magnitudes of about -6.5. This, coupled with the observed magnitudes, allows us to make an estimate of the distance of that galaxy. We must, of course, be aware that this method of distance determination could be inaccurate, but it does give the order of distance.

Another method of determining the distance of the Andromeda galaxy, using the decay rate of novae, is given in the next project.

8.9   PROJECT 54

To determine the distance of the Andromeda galaxy from the rate of decay of brightness of novae observed in it.

Some years ago, H. C. Arp, of Mount Wilson and Palomar Observatories, published a study of many novae which have appeared in the Andromeda galaxy. The apparent magnitudes of the novae at their brightest were plotted against the observed rate of fading, showing a relation between these two quantities. Since the Andromeda galaxy is so far away from us, all these novae can be considered to be at the same distance from us. It follows that there is a similar relation between the absolute magnitudes of these novae at maximum and their rate of decay of brightness.

Furthermore, novae in our own Galaxy, the Milky Way, follow a similar relation. Since many of these novae are close enough for their distances to be determined by the method of parallax, it is possible to plot absolute magnitudes at maximum against rates of decay.

If then the two graphs are moved parallel to the magnitude axes so that they eventually coincide, the distance moved will be the difference between the absolute and apparent magnitudes of the novae in the Andromeda galaxy.

The parallax of these novae will be given by the formula

$$M = m + 5 + 5 \log_{10} p$$

and the distance in parsecs will be given by $1/p$.

The following Tables give the absolute magnitude at maximum of a number of novae in the Milky Way and their observed rate of decline in magnitude, and similarly the apparent magnitudes at maximum of a number of novae in the Andromeda galaxy and the corresponding observed rates of decay of magnitude.

Set up axes of rate of decline of the novae in the Andromeda galaxy as abscissa to a scale of   20 mm = 0.1 magnitudes   per day, and apparent magnitude of these novae as ordinate to a scale of   50 mm = 1 magnitude. Label the abscissa 0 at the intersection of the axes and extend the scale as far as 0.7. Label the ordinate axis 18 at the intersection of the axes and extend this scale upwards as far as the value 15. The values will decrease numerically upwards since the smaller the number the brighter the nova. Plot the values given in the Table for the Andromeda galaxy.

On a piece of tracing paper set up axes of rate of decline against absolute magnitude to the same scales as before, but label the abscissa 0 to 0.7, and the ordinate − 6 at the intersection of the axes upwards to − 9 for the absolute magnitudes. Plot, on these axes, the values given in the Table for the Milky Way.

The values for the Milky Way give a well defined graph which should be drawn in. Due to the larger distances at which the novae in the Andromeda galaxy have been observed, the points for this graph are much more scattered, but it is possible to draw in lines similar to those for the Milky Way. As many points as possible should be included.

The graph for the Andromeda galaxy should now be placed under the

TABLE 8.3 Novae in the Milky Way

Absolute magnitude at maximum	Rate of decline (magnitudes per day)
−8.4	0.67
−8.4	0.52
−8.35	0.57
−8.35	0.50
−8.3	0.40
−8.3	0.33
−8.3	0.30
−8.3	0.29
−8.25	0.25
−8.25	0.20
−8.25	0.17
−8.1	0.15
−7.95	0.14
−7.8	0.13
−7.7	0.12
−7.55	0.118
−7.35	0.105
−7.3	0.100
−7.15	0.091
−7.05	0.083
−6.75	0.064
−6.65	0.054
−6.6	0.050
−6.5	0.045
−6.45	0.040
−6.4	0.034
−6.4	0.032
−6.25	0.030
−6.2	0.027

tracing paper which carries the graph for the Milky Way and the two sheets adjusted so that the two graph lines lie on top of one another.

Read any two corresponding values on the two magnitude axes and subtract them, which, since one value will be negative, amounts to adding the values numerically. This will give $M - m$ in the formula

$$M - m = 5 + 5 \log_{10} p$$

TABLE 8.4 Novae in the Andromeda galaxy

Apparent magnitude at maximum	Rate of decline (magnitudes per day)
15.9	0.38
18.2	0.20
15.9	0.17
16.0	0.23
15.9	0.15
16.0	0.23
16.0	0.174
16.0	0.29
16.1	0.180
17.0	0.077
16.2	0.163
16.4	0.126
16.7	0.156
17.2	0.069
17.5	0.058
17.6	0.070
17.2	0.060
17.4	0.075
17.6	0.067
17.4	0.046
17.8	0.059
17.6	0.061
18.0	0.043

where, as before, M = absolute magnitude of the nova
m = apparent magnitude of the nova
p = parallax of nova in seconds of arc

Calculate the parallax p and the distance of the novae in the Andromeda galaxy $1/p$ in parsecs. Multiply this by 3.26 to obtain the distance in light years. Compare the value obtained with the value 2×10^6 light years.

It is interesting also to compare these values for distance with that obtained by assuming that the absolute magnitude of an average nova in the Andromeda galaxy is about -6.5. For the apparent magnitude take a value midway between the extremes in the Andromeda galaxy table.

8.10   INTERSTELLAR MATTER

The space between the stars is not empty. It is thought that originally space contained gaseous hydrogen from which the stars were formed. The pressures and temperatures created by the formation of the stars through gravitational attraction were sufficient for nuclear reactions to begin within the stars, during which hydrogen was converted into helium and other heavier elements. Some of the stars have gone through the processes of evolution, and have eventually exploded, throwing out into interstellar space the elements manufactured during their lives. Thus space contains hydrogen and dust formed from other elements.

For our own Galaxy it has been estimated that about one tenth of the total mass within the Galaxy is interstellar dust and gas. The rest is concentrated in the stars. The interstellar matter itself is about one per cent dust and ninety-nine per cent gas. Observations have shown the presence of those elements and compounds already mentioned in the introduction.

We also know that there are regions of interstellar space in which there are concentrations of dust, or grains, as they are sometimes called. The size of these grains is not known since the grains cannot be measured directly, but from some of the effects they produce on the light passing from stars through the dust clouds, and from consideration of the properties of grains of materials likely to be found in interstellar space, it is thought that they are of the order of 0.0001 mm.

Evidence of interstellar dust and gas is provided by patches of nebulosity which obtain their illumination from nearby stars, and are observed as hazy areas of light. Often in these lighter patches there are dark patches where the concentration of interstellar dust is relatively high. Further evidence is provided by the reddening of the light from stars due to passing through interstellar grain concentrations, and also by an effect on this light known as polarisation.

8.11   THE DARK NEBULAE

Figure 8.8 is a photograph of the North America nebula, so called for its remarkable resemblance to the outline map of that continent. We may note that in the region corresponding to the Gulf of Mexico and east of Cape Canaveral there is a concentrated dark area which seems to be relatively devoid of stars when compared with the nebula itself.

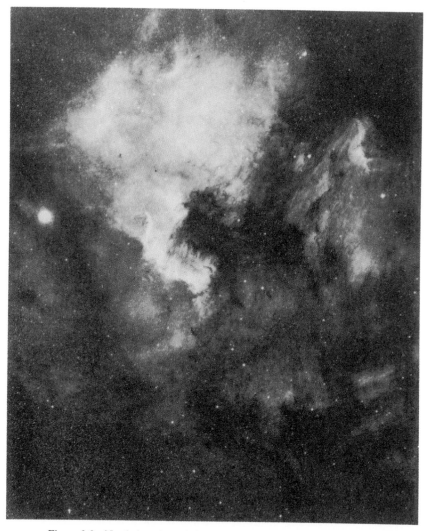

Figure 8.8   North America nebula   *Photo © National Geographic Society, Palamar*
Sky Survey
Reproduced by kind permission

One reason why this could be so is that between the observer and the
stars in that region there is a dense cloud of interstellar dust which is
progressively reducing the apparent brightness of stars within the cloud
as the distance from the observer increases and of reducing the brightness of, or even completely obscuring, those stars which lie behind the

cloud. The reduction in brightness may be of the order of several magnitudes when compared with close-by stellar areas which are relatively free from this interference.

We may draw a Wolf diagram as in Figure 8.9 in which the logarithm of the number of stars (log N) of apparent magnitude m is plotted against

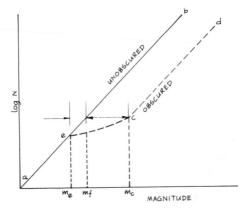

Figure 8.9   Interstellar cloud quantities

m. Line ab represents the result which one would expect in unit area of relatively unobserved sky. Line cd represents the result in an equal area of the observed region. We note that the number of stars of a given magnitude is less for cd than for ab. The indeterminate line ec represents the transition from unobscured stars in front of the cloud through the gradual diminution of brightness through the cloud.

If we assume a mean absolute brightness of M for stars at this distance then we may apply

$$M = m + 5 + 5 \log p''$$

to obtain three pieces of information about the dark nebula.

(a) Given M and m_e (Figure 8.9) which is the apparent magnitude at which the interstellar material starts its interference, we may calculate p''_e and hence the distance of the dark nebula.

(b) Given M and $(m_f - m_e)$ which is the change in apparent magnitude due to the whole cloud, we may deduce the extent of the cloud in the line of sight by calculating say p''_{f-e}.

(c) Given $(m_c - m_f)$ we may deduce the number of magnitudes which are due to the absorption of the whole cloud.

8.12   PROJECT 55

(i)   To determine the distance of the North America nebula, using information relating to an area of dark nebulosity in that nebula.

(ii)   To determine the total absorption, that is the reduction in magnitude of starlight, due to the whole cloud of interstellar matter.

(iii)   To determine the depth, in the line of sight, of the cloud.

Data

Figure 8.10 shows the North America nebula again but this time as a negative print in which the bright stars appear as dark spots. It is easier to take measurements from this type of print.

On the photograph is marked an area of approximately one square degree which covers part of the cloud of interstellar matter.

The Table below gives star counts in a nearby unobserved region of the sky.

TABLE 8.5

Apparent magnitude of star	Number of stars per square degree of sky
8.0	1.12
9.0	3.31
10.0	9.33
11.0	26.92
12.0	77.62
13.0	199.5
14.0	542.8
15.0	1318
16.0	2512

For this project we shall need a device for measuring the diameter of star images on the photograph to the nearest one-tenth of a millimetre. This will mean a microscope fitted with a graticule or magnifying eyepiece similarly fitted. The author is aware that this deviates from the policy of the book that only the simplest of equipment is needed, but it was thought worthwhile to include this project for those who have access to the more sophisticated measuring equipment.

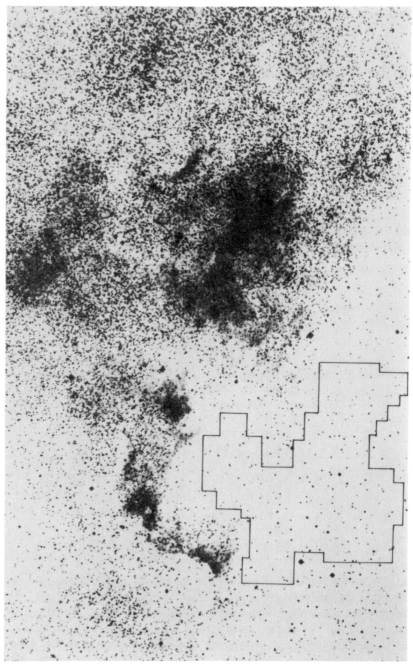

Figure 8.10   Location of area of interstellar matter   *Based on an original photograph* ©
National Geographic Society, Palamar
Sky Survey
Reproduced by kind permission

Divide the area marked out in the photograph into convenient rectangles. Measure the diameter of each star. Estimate these to the nearest one-tenth of a millimetre. The greatest difficulty in doing this is keeping track of the stars already measured. This is tedious, but much of the work of the professional astronomer is spent, not at the eyepiece of a telescope, but in analysing results such as these.

It is now necessary to decide what the diameter of the image of a star means in terms of apparent magnitude. If Project 39 has already been carried out we shall already have a graph relating diameter and magnitude. If Project 39 has not been carried out a single graph only, that of magnitude as ordinate against star image diameter as abscissa, needs to be drawn, using the data given for that project.

From the calibration curve read off the apparent magnitude of each of the stars previously measured on the North America nebula photograph. Count the number of stars whose magnitude falls in the range 8.5 to 9.4 inclusive, to be counted as magnitude 9. Do a similar count for 9.5 to 10.4 (magnitude 10) and so on up to and including magnitude 15.

Now plot a graph of $\log_{10} N$ as ordinate, where N is the number of stars per square degree of sky counted of magnitude m, against m the magnitude. For the $\log_{10} N$ axis use a scale of 50 mm = 1 unit, and for the m axis use a scale of 10 mm = 1 magnitude. This will give the part of the graph ecd and possibly a little of ae as shown in Figure 8.9.

The unobscured part of the graph aeb can be obtained by plotting the values given in the Table on the same graph sheet, noting that we are still plotting $\log_{10} N$.

Read off the value of m_e (Figure 8.9). This corresponds to the start of obscuration, and if we assume an average absolute magnitude of stars in the nebula, then we can obtain an estimate for the distance of the nebula from the now well used formula

$$M = m + 5 + 5 \log_{10} p \qquad \text{[i]}$$

where $M = 5$ (assumed)
$m = m_e$ (the value read off the graph)
$p = $ the parallax in seconds of arc.

From this formula find p. The distance of the nebula is then given by $1/p$.

Compare the value obtained with 300 parsecs.

Next measure the quantity $m_c - m_f$ (Figure 8.9). This represents the number of magnitudes by which all stars in the region behind the dark nebula are reduced in brightness. It is therefore the total absorption of the cloud of interstellar matter.

Finally, read off the quantity $m_f - m_e$ (Figure 8.9), which is the magnitude range on the unobscured scale over which the cloud is effective. We can obtain the distance necessary for the magnitude of a star to fall by $m_f - m_e$ from

$$m_f - m_e = 5 + 5 \log_{10} p_{f-e}$$

an adaption of formula [i] above. Work out p_{f-e} and also $1/p_{f-e}$, the distance in parsecs corresponding to the thickness of the cloud. Divide the quantity $m_c - m_f$ by the quantity $1/p_{f-e}$ to give the obscuration in magnitudes per parsec. Divide by 1000 to obtain the obscuration in magnitudes per kiloparsec.

Compare this with the value of 1.5 magnitudes per kiloparsec, which is roughly an average for interstellar absorption, although this may vary locally from about 0.5 to 4 magnitudes per kiloparsec. Check that the value obtained from this project is reasonable according to the figures just quoted.

8.13   THE FORM OF GALAXIES

Galaxies have been classified by Hubble and Sandage from those with a spherical form, through elliptical forms to spirals, like our own Galaxy and the Andromeda galaxy M31, and barred spirals. The shape of the arms in spiral galaxies often approximate closely to logarithmic spirals. A property of this type of spiral is that lines drawn from the beginning of the spiral all make the same angle with the related tangent at the point where the line cuts the spiral curve.

8.14   PROJECT 56

To check whether or not the spiral arms in the galaxy M74, type Sc, are in fact logarithmic spirals.

Trace the spiral arm ab on the photograph of M74 (Figure 8.11) from the point a where the spiral arm leaves the core of the galaxy to the furthermost visible point b.

Figure 8.11   Spiral galaxy M74   *Photo © Hale Observatories, Pasadena, U.S.A.*
Reproduced by kind permission

From *a* draw several radii well spaced to cut the spiral (Figure 8.12). At each of the intersection points draw the tangent to the spiral by eye, and measure the angle θ between the radii and corresponding tangents.

Compare the values with 70° to 75° which is a typical range for this type of galaxy.

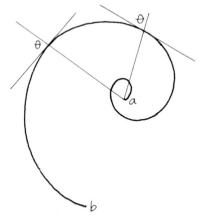

Figure 8.12   Spiral arm

8.15   HUBBLE'S LAW

Using the method of distance determination by observation of the Cepheid variables (see Project 53), Edwin Hubble was able to measure the distances of some nebulae, which consist really of a very large number of stars as does our own Galaxy. He later found the remarkable relation mentioned in a preceding section, namely that the distance of such objects was proportional to their velocity of recession, which is related to the red shift (section 7.30).

The next project is concerned with demonstrating that this is so, and the three following projects illustrate three different cases of how the application of the phenomenon of the red shift observed in spectra can be profitably used.

8.16   PROJECT 57

(i)   To plot the velocity of recession of a number of galaxies against the distance of these galaxies, and to determine Hubble's constant.

(ii) To arrive at an estimate for the distance to the edge of the observable universe, assuming Hubble's law holds good for such distances.

Data

The figures below refer to the velocity of recession of a number of galaxies and the corresponding distances. The galaxies are referred to by the name of the constellation of stars in which they are found.

TABLE 8.6

Constellation of nebula	Distance (light years)	Velocity of recession (km s^{-1})
Virgo (the Virgin)	39×10^6	1 200
Ursa Major (the Great Bear)	485×10^6	14 950
Corona Borealis (the Northern Crown)	700×10^6	21 600
Boötes (the Herdsman)	1280×10^6	39 200

Plot the velocity of recession as abscissa to a scale of 10 mm = 2000 km s^{-1} against distance as ordinate to a scale of 10 mm = 100 $\times$ 10^6 light years. Draw a straight line through the origin and the points on the graph. Note that the points *do* lie on a straight line through the origin, indicating that the velocity of recession is proportional to distance.

Take the slope of the line, and express this in km s^{-1} per megaparsec of distance. One megaparsec is one million parsecs, and we shall also need to recall that 1 parsec = 3.26 light years.

Compare your value with a typical estimate of Hubble's constant, namely 100 km s^{-1} per megaparsec. There is an uncertainty in the true value of this, values between 50 and 125 km s^{-1} per mpc being acceptable.

Given that a galaxy in the constellation of Hydra (the Sea-serpent) has a velocity of recession of 60 800 km s^{-1}, estimate its distance in light years and compare this with the value 4500×10^6 light years.

Hubble's law can thus be expressed in the form

$$v = 100d$$

where v = velocity of recession in km s^{-1}

d = distance in megaparsecs

When the velocity of an object, such as a nebula, is receding at the speed of light $(3 \times 10^5$ km s$^{-1})$, the light waves would not reach us, and we should be unable to see it, even with the more powerful telescopes which might yet be devised. Assuming that Hubble's law applies to such great distances and velocities, insert the velocity of light for v in the above equation, and thereby determine the distance of the observable universe.

Compare this with 10 000 million light years. At the time of writing objects as far away as 5000 million light years have been observed.

8.17 QUASARS

Quasar is a name coined from quasi-stellar objects. These objects, discovered as recently as the early 1960s, have proved to be some of the most difficult objects to explain in the whole universe.

Basically their chief characteristic is that they are powerful transmitters of radiation, usually on those wavelengths which can be detected by radiotelescopes. They are also very small in size, so that their size and energy output do not seem compatible compared with other, more well known, astronomical objects. Some of the radio sources have been identified with optical objects which can be photographed using large telescopes. Some appear as star-like objects and others as nebulae, all of which radiate strongly in the ultraviolet region of the spectrum.

One of the most puzzling things about quasars is that they show very large red shifts in their spectra. If red shifts can be converted into velocities, then the velocities of some quasars are very high indeed, as we shall see in the next project. Also, if such high velocities can be converted into distances as indicated by Hubble's law, then some of the quasars are at an immense distance, which does not seem to agree with the large radiation which we receive from them.

Some astronomers have tried to explain the large velocities, as indicated by red shifts, as being those of objects much nearer, in our own Galaxy perhaps, which have been flung outwards by a great explosion sometime in the past. One objection to this is that one would expect at least some blue shifts also from such an explosion in our Galaxy, but none have been detected.

At the time of writing, the red shift as a measure of velocity and great distance seems to find greater favour amongst astronomers than any other explanation.

8.18   PROJECT 58

To determine the red shift of light from a quasar, and, assuming that the relationship between red shift and velocity holds for such large red shifts, to determine the velocity of recession of the quasar.

Data

Wavelengths of elements at rest:

Silicon II	1194 Angstrom units
Silicon III	1207
Lyman a (Hydrogen)	1216
Nitrogen V	1240
Silicon II	1263
Carbon II	1335
Silicon IV	1395
Silicon IV	1403
Carbon IV	1549
Helium II	1640

The red shift for this quasar 3C 191, is so large that the whole of the spectrum covering the rest wavelength and the red shift wavelength cannot be shown to a scale on a paper of this size. Consequently the rest wavelengths are given above, and the part of the spectrum showing the displaced lines for the corresponding elements only is shown on the left of Figure 8.13, alongside a calibration spectrum which extends over the range of wavelength of the displaced lines.

We are not entitled to assume that the spectrum is linear with wavelength over such a range, and so we shall first draw a calibration graph from the right hand spectrum.

About 20 mm above the Mercury 3651 line on the calibration spectrum, draw a horizontal line at right angles to each spectrum and extending across them both. Take distances in millimetres on the calibration spectrum of each line from the datum line just drawn. Plot these distances as ordinate to a scale of 10 mm = 10 mm, and beginning at zero, against wavelength of the corresponding lines as abscissa to a scale of 10 mm = 100 Å, beginning at 3500 Å. Draw in the best curve to fit these points.

Next, measure in millimetres the displacement of the element lines in the

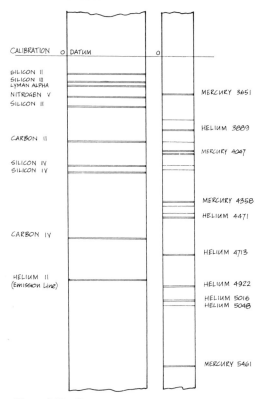

Figure 8.13 Quasar spectrum

Line illustration based on an original spectra print as devised by Burbidge & Holye which in their text "Frontiers in Astronomy". © *The Kitt Peak National Observatory, Arizona, U.S.A.*

displaced spectrum from the datum line. Read off from the calibration graph the displaced wavelength of each of these elements. Subtract the wavelength at rest from the displaced wavelength, and divide the result by the rest wavelength to give a quantity, say X, for each element. Calculate the average value of X.

Since the red shift is so large that we might expect a velocity of recession approaching the velocity of light, the formula for calculating the velocity will differ from that used in Project 47. The formula to be used in cases such as this is

$$1+X = \sqrt{\frac{1+\dfrac{v}{c}}{1-\dfrac{v}{c}}}$$

where

v = the velocity of recession
c = the velocity of light
= 3×10^5 km s^{-1}.

By squaring both sides of this equation, and re-arranging the quantities, determine the quantity v/c.

Compare this with the value 0.8, which indicates that the speed of recession of this quasar is about 80% of the speed of light.

8.19   GALACTIC CO–ORDINATES

In this book we have travelled almost to the edge of the observable universe. We have used the spectra of distant galaxies to do this and we now apply our knowledge of specific spectra to our own Galaxy. Section 8.1 describes our Galaxy as containing most of its stars in a disc, the Milky Way, and it is sometimes convenient to use this disc as a reference plane when quoting the positions of objects within our own Galaxy. This then leads us to define a fourth system of co-ordinates known as galactic co-ordinates. As with the three systems described in Section 4.2 these can be likened to latitude and longitude on Earth. The galactic equator lies within the disc of stars which passes through the centre of the Galaxy. In the Ohlsson system galactic latitudes (b) are measured from this equator, north and south, while longitudes (l) are measured from one of the two points, or nodes, where the galactic equator cuts the celestial equator (Section 4.1). The point from which galactic longitudes are measured is that node at which declinations change from negative to positive (the ascending node). In Figure 8.14 this point $l = 0°$ is denoted by 0.

To return to the spectra, it is a feature of Type O stars that they display bright, or emission lines. Indeed, some of the Type O stars seem to have

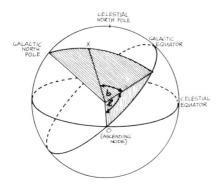

Figure 8.14   Galactic co-ordinates

a temperature even greater than that quoted in Table 7.5 and these have been given the spectral classification Type W, after Wolf (and Rayet) the two French astronomers who discovered them.

8.20   PROJECT 59

To determine the right ascension and declination of reference points for galactic longitude l and galactic latitude b, namely the equatorial co-ordinates for $l = 0°$ from which longitudes are measured and for $b = +90°$ which defines the north pole of the Galaxy.

Data

The Table below gives the equatorial co-ordinates of some Type O stars which are known to belong to our own Galaxy. They are taken from the Henry Draper Catalogue.

It is known from observation that Wolf-Rayet stars are usually found near to the Galactic equator. By plotting their positions in terms of right ascension and declination a curve can be drawn through the distribution. The curve drawn will correspond to the Galactic equator.

Taking declination as ordinate and using a scale of 10 mm = 10° of declination, draw this axis close to a short edge of the paper and parallel to it. From the centre of this axis mark the declination scale from 0° to 80° upwards and 0° to −80° downwards. Through zero declination draw the right ascension axis as abscissa parallel to the longer sides of

Star Number	α		δ		Star Number	α		δ	
53667	7h	01m	$-8°$	34'	97152	11h	06m	$-60°$	26'
54662	7	05	-10	11	97253	11	07	-59	50
56925	7	14	-13	03	97434	11	08	-60	09
57060	7	15	-24	23	97950	11	11	-60	43
57061	7	15	-24	47	104994	12	00	-61	29
60848	7	31	$+17$	07	105056	12	01	-69	01
62150	7	37	-32	24	112244	12	50	-56	17
62910	7	41	-31	41	113904	13	02	-64	46
63099	7	42	-34	05	115473	13	12	-57	37
63150	7	42	-36	16	117297	13	24	-61	34
65865	7	56	-28	28	117688	13	27	-61	48
66811	8	00	-39	43	117797	13	28	-61	54
68273	8	07	-47	03	119078	13	36	-66	54
69106	8	10	-36	38	120521	13	45	-58	03
73882	8	36	-40	04	121194	13	49	-60	40
76536	8	52	-47	13	124314	14	08	-61	14
79573	9	10	-49	42	134877	15	07	-59	28
86161	9	52	-52	15	135240	15	09	-60	35
88500	10	07	-60	09	135591	15	11	-60	08
89358	10	14	-57	25	136488	15	16	-62	19
90657	10	23	-58	08	137603	15	22	-58	14
91824	10	31	-57	39	143414	15	55	-62	24
91969	10	32	-57	43	147419	16	17	-51	18
92554	10	36	-60	24	149038	16	27	-43	50
92740	10	37	-59	09	150135	16	34	-48	34
92809	10	38	-58	15	150958	16	39	-46	55
93128	10	40	-59	02	151804	16	45	-41	04
93131	10	40	-59	36	151932	16	45	-41	41
93162	10	40	-59	12	152147	16	47	-41	57
93250	10	41	-59	03	152233	16	47	-41	37
93843	10	45	-59	42	152270	16	47	-41	40
94305	10	48	-61	46	152386	16	48	-44	50
94546	10	50	-58	59	152408	16	48	-41	00
94663	10	51	-58	16	152424	16	48	-41	56
95435	10	56	-57	17	153919	16	57	-37	42
96548	11	02	-64	58	156327	17	12	-34	18

Star Number	α		δ		Star Number	α		δ	
156385	17h	12m	−45°	32′	184738	19h	31m	+30°	18′
157451	17	16	−43	24	186943	19	42	+28	01
157504	17	19	−34	06	187181	19	44	+17	57
158860	17	27	−33	33	188001	19	48	+18	25
159176	17	28	−32	31	190002	19	58	+32	18
160529	17	35	−33	27	190429	20	00	+35	45
163181	17	50	−32	27	190864	20	02	+35	19
163454	17	51	−31	00	190918	20	02	+35	31
163758	17	53	−36	00	191765	20	07	+35	53
164270	17	55	−32	43	191899	20	07	+11	35
164492	17	56	−23	01	192103	20	08	+35	54
164794	17	58	−24	22	192163	20	08	+38	03
165052	17	59	−24	24	192639	20	11	+37	03
165688	18	02	−19	25	192641	20	11	+36	21
165763	18	03	−21	16	193077	20	13	+37	07
166813	18	07	−42	53	193576	20	16	+38	25
167264	18	09	−20	46	193793	20	17	+43	32
167633	18	11	−16	33	193928	20	18	+36	36
167771	18	12	−18	30	195177	20	25	+38	17
168206	18	14	−11	40	199579	20	53	+44	33
169010	18	18	−13	46	203064	21	15	+43	31
175876	18	52	−20	33	206267	21	36	+57	02
177230	18	59	−4	28	208220	21	50	+43	01

the paper. Mark the right ascension scale from 0 hours to 24 hours using a scale of   10° mm = 1 hour of right ascension.

Plot the position of each star given in the Table. Draw in the best curve through as many points as possible leaving out the very few which clearly deviate from the normal trend.

Read off the right ascension of the point where the trace of the Galactic equator just drawn cuts the right ascension axis as it moves from negative through zero to positive values of declination. This is the reference point $l = 0°$ from which galactic longitudes are measured.

Next read off the right ascension and declination of the point where the curve reaches its minimum value in declination. At this point the great circle which cuts the galactic equator at right angles will pass through the galactic north pole. The galactic north pole will therefore be 90° in declination away from this point. The north pole of the Galaxy has therefore equatorial co-ordinates equal to the right ascension just read and declination 90° plus the declination just determined.

Compare the values obtained from the graph with $a = 18$h 40m for the reference point $l = 0°$, and $a = 12$h 40m, $\delta = +28°$ for the north pole of the galaxy.

9

Telescopes

9.1 TELESCOPES

The projects and the accompanying notes in this book cover many of the topics of interest to the astronomer, both amateur and professional. It would not be proper to conclude without a few comments on the one instrument which has assisted most in the collection of such information, namely the telescope.

It is thought that Galileo was the first person to turn one of the newly-invented telescopes to the sky. Compared with the precision instruments of today, Galileo's telescope must have been difficult to use, but nevertheless he saw the craters on the Moon and four of the moons of Jupiter. His telescope was of the refracting type (see Figure 9.1), consisting of a large diameter lens O, of long focal length and known as the objective, which is used to collect light, and, by refraction, to bring the rays to a focus and form an image. This image can then be magnified by the use of a second lens or lens system E, of short focal length and known as the eye-piece, at the other end of the tube. Arrangement is made to vary the distance between the objective and eyepiece to take account of the different distances of the objects which may be viewed. Galileo's eyepiece was a concave lens, but telescopes nowadays usually have convex lenses in the eyepiece.

Such a telescope is limited in size by the difficulties of making large objective lenses. It also has a number of defects, some of which can be corrected, and others which cannot. One of these defects is known as chromatic aberration, which is due to the objective lens behaving like a prism, and splitting up the white light into its component colours of the rainbow, so that the image is confused by coloration. This defect can mostly be overcome by making the objective of two lenses, one convex made of crown glass, and the other concave made of flint glass, the two

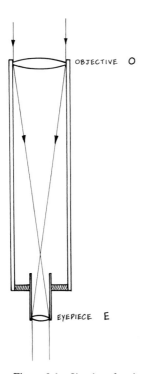

OBJECTIVE   O

EYEPIECE   E

Figure 9.1   Simple refracting telescope

types of glass having different refractive indices, that is, having different refractive properties.

Another defect of the refracting telescope is that due to the spherical nature of the lens surfaces. Those rays which are near to the axis of the telescope are brought to a focus at a point further from the lens than are those some distance from the telescope axis. This effect is known as spherical aberration. The image in this case is spread out, both along the axis and at right angles to the axis, to form what is known as a caustic curve. The so-called circle of least confusion is formed five-eighths of the distance from the focus furthest from the lens to the focus nearest the lens. Project 60 shows how this defect is formed, but takes a spherical mirror instead of a spherical lens, since the rays of light are easier to construct.

A second and most important type of telescope is the reflecting telescope, one kind of which was designed by Isaac Newton (see Figure 9.2).

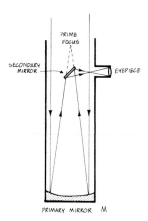

Figure 9.2   Newtonian reflecting telescope

The collection of light is carried out by a mirror M whose front reflecting surface is made into the shape of a paraboloid, which has the property of bringing all rays parallel to the axis of the telescope to a sharp focus, known as the prime focus. The mirror of the Hale telescope at Mount Palomar is very large, 200 inches, or over 5000 mm, in diameter, so that it is possible to put an observer's cage, large enough to take the observer and his camera, near to the prime focus without too much loss of light due to the obstruction of the cage. The new Isaac Newton telescope at the Royal Greenwich Observatory at Herstmonceux, in Sussex, has a similar cage at the prime focus, the diameter of the main mirror in this case being 98 inches, or about 2500 mm. With smaller telescopes, it is usual to put a secondary mirror, which, in the case of a Newtonian reflector, is plane, and which deflects the light to a focus at the side of the telescope, near to the top end of the tube. Because of the simplicity of the optics, this type is much favoured by amateur astronomers. It is not very convenient, however, to have an eyepiece some distance from the ground, and whose position in space changes with the direction of the telescope.

For this reason the Cassegrain (see Figure 9.3) and the Gregorian (see Figure 9.4) types of telescope were invented.

The main mirror in each of these types has a central hole. In the case of a Cassegrain telescope the light collected by the main mirror is allowed to strike a secondary mirror, which has a convex hyperbolic surface, which in turn directs the light through the hole in the main mirror to form a focus behind the main mirror. Thus the eyepiece is in the same

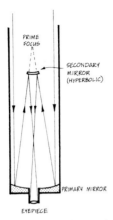

Figure 9.3   Cassegrain reflecting telescope

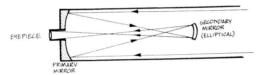

Figure 9.4   Gregorian reflecting telescope

position relative to the tube as a refractor, and observation is always along the axis of the tube, a much more convenient process. Also due to the doubling of the length of the ray paths, the Cassegrain telescope is compact.

The Gregorian type of telescope has an elliptical concave secondary mirror which again brings the rays to a focus at the back of the main mirror.

There are other types of telescope, some using slightly different principles, but most of the others being variations of the types described above.

We began this book by a study of the relevant parts of the properties of conic sections, and we approach the end of the book using the conic sections in a quite different way.

The reflecting telescope suffers no chromatic aberration, since the light is reflected at the front surface of the mirror and therefore does not pass through the glass in which most of the chromatic aberration occurs. It can, however, suffer a distortion of the image known as coma, which can arise from the rays approaching the telescope at an angle to the optical axis. The appearance of coma in a telescope is like a teardrop, round at one end and tapering to a point at the other. Project 60 has been designed to show the formation of coma even from a correctly shaped parabolic mirror.

Astronomers using optical aids such as the telescope are continually trying to increase the distance at which they can observe light radiation. As with the radio-astronomer (see Section 9.6) the brightnesses of the objects observed might be very much less than the faint background light which covers the sky. The integration which takes place on the photographic plates by using long exposures applies also to the background, and there is a saturation limit. The use of larger telescopes might go some way to solving the problem but unfortunately the cost of telescopes increases at a much greater rate than does the light-gathering power.

Today the astronomer has several devices to assist him with this signal-to-noise ratio problem. One of these is the photomultiplier in which photons falling on a cathode may release a larger number of electrons. The latter are then accelerated by an electric field until they strike another element, the dynode. Here each electron releases a number of other electrons which in turn are accelerated. The process can be repeated several times with a consequent amplification of about 10^6. The baseplate on which the final electrons fall builds up an image of the object being observed as a distribution of electric charge which can be scanned as the picture in a television set is scanned. The intensities of the parts of the image can be fed into the memory addresses of a computer, so that a continually intensifying picture can be reproduced at any time, even while the image is being built up.

9.2   THE NORMAL TO A PARABOLA

It is a well known optical law that a ray of light striking a reflecting surface will be reflected such that the angle which the reflected ray makes with the normal is equal to the angle which the incident ray makes with the normal to the surface at the point of contact.

With a plane mirror the normal is easy to construct, being at right angles to the reflecting surface of the mirror (Figure 9.5). With a spherical mirror the normal is also easy to construct, being the radius from the point of contact to the centre of curvature of the spherical surface (Figure 9.6). In the case of reflection from a parabolic surface, however, the construction is not so obvious, but is equally simple if the position of the focus of the parabola is known.

Referring to Figure 9.7, P is a point on the parabola at which the

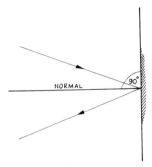

Figure 9.5   Normal to plane

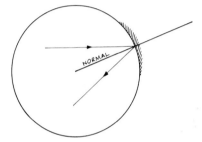

Figure 9.6   Normal to sphere

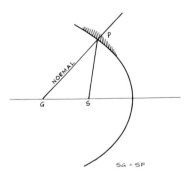

Figure 9.7   Normal to parabola

normal is to be drawn, and S is the focus of the parabola. Suppose the normal, when drawn, cuts the axis of the parabola in G. Then it is a property of the parabola that $SG = SP$. Thus, to construct the normal at the point P it is only necessary to mark off SG equal to SP, and join PG which will be the normal required.

9.3   EQUAL ANGLE CONSTRUCTION

In both Project 60 and Project 61, after constructing the normal to the reflecting surface it will be necessary to make the angle of reflection equal to the angle of incidence of a ray of light.

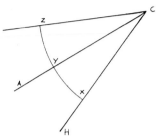

Figure 9.8   Equal angle construction

In Figure 9.8, let *HC* be the incident ray and *CA* the normal to the mirror. With centre *C* and any reasonable radius, draw an arc to cut *HC* in *X* and *CA* in *Y*. Using compasses or dividers, mark off a point *Z* on the arc *XY* produced such that *YZ* = *XY*. Join *CZ*. Then *CZ* is the reflected ray, and ∠ *HCA* = ∠ *ACZ*.

9.4   PROJECT 60

To demonstrate how spherical aberration arises, in this case on reflection at a spherical mirror, and to show how the caustic curve is formed.

Data

We shall assume that the mirror has a radius of 200 mm, so that the aberration will be sufficiently emphasised. In practice no mirror of such a radius would be used at such a large aperture as in this project.

With the paper longways on, draw the central horizontal line to represent the axis of the mirror. Locate the centre of curvature of the mirror *O* near to the right hand end of this axis. With centre *O* and radius 200 mm draw an arc of a circle extending about 80 mm on each side of the axis to represent the reflecting surface of the mirror.

Draw horizontal lines *DZ*, *CY*, *BX* and *AW* on one side of the axis and respectively 20, 40, 60 and 80 mm from it to represent rays from a distant object striking the mirror at points *Z*, *Y*, *X* and *W*. Do likewise with four rays *D'Z'*, *C'Y'*, *B'X'* and *A'W'* on the other side of the axis.

From each of the points W, X, Y and Z and W', X', Y' and Z' draw a line to the centre O. These lines will be the normals to the surface at these points. Using the construction described in Section 9.3 draw from W the line which makes the same angle with the normal at that point as does the incident ray AW. This line will be the reflected ray cutting the axis at, say A'. Do likewise for all the other incident rays cutting the axis in points B', C' and D'.

It will be clear from the drawing that rays near to the axis give reflected rays coming to a focus further from the mirror than those rays further from the axis. This is spherical aberration, clear focusing being impossible between A' and D'. The so-called circle of least confusion is located five-eighths of the way from D' to A'.

Note also that the reflected rays outline a curve near to the various foci, and of course symmetrical about the axis, which is the caustic curve.

9.5   PROJECT 61

(i)   To demonstrate that all rays parallel to the axis of a parabolic mirror are reflected through the focus of the parabola.

(ii)   To demonstrate the effect known as coma in a parabolic mirror due to a set of non-axial parallel rays.

Data

For this project, the curvature of the mirror will be greatly exaggerated so that the coma produced will be more obvious. The diameter of the mirror selected for the project is 200 mm, and the depth of the mirror at the centre 25 mm. In an actual telescope the depth at the centre of a mirror of this diameter would be of the order of 2 mm. The focal length of the mirror in the project is 100 mm, whereas in an actual telescope it would probably be about 1600 mm. The effect of coma is increased with increase in the curvature of such mirrors.

With the paper longways on, set up a vertical axis 15 mm from the right hand edge of the paper, and a horizontal axis, to represent the axis of the mirror, about 60 mm from the bottom of the paper. Label the point of intersection of the axes O, and measure OB 100 mm vertically from O. Measure BC to the left of B a distance of 25 mm. Then C is a point on the parabolic surface. At intervals of 20 mm below B, locate points D,

E, F and *G* respectively 16, 9, 4 and 1 mm from the vertical *OB*. Then *CDEFGO* is the reflecting surface for half the mirror. It may be helpful to join these points by a faint smooth curve, although this is not absolutely necessary since we can construct the normals at these points without reference to the curve itself.

Measure *OS* to the left of *O* a distance of 100 mm. *S* will then be the focus of the mirror. Measure from *S* to the left a distance *SP = SC*. Join *PC* which will be the normal to the curve of the mirror at *C*. Take a horizontal ray striking the mirror at *C*. Construct on the other side of the normal an angle to that between the incident ray and the normal. This will be the reflected ray.

Similarly construct the reflected rays at *D, E, F* and *G*, and show that they all pass through the focus *S*.

On a separate piece of paper, construct the mirror surface as before, and also the normals at each of the points. Taking incident rays through these points, but arriving at these points from a direction 30 degrees below the horizontal, construct the reflected rays at each of the points *C, D, E, F, G* and *O*.

Observe that these reflected rays do not meet in any one point. By drawing through *S*, the focal plane, show that the scatter of the reflected rays is widespread, giving the coma effect when a distant object is observed through the telescope eyepiece.

9.6   RADIO-ASTRONOMY

The science of radio-astronomy is very young when compared with the time that optical telescopes have been used. Many celestial objects, galaxies, stars and even planets radiate at wavelengths comparable with those of radio waves. This radiation can be received by antennae or aerials in much the same way as by those for domestic television and radio and we shall all be familiar with the parabolic dish type of aerial such as the one at Jodrell Bank, in England. The strength of the radiation is very small indeed. In fact, it may have characteristics similar to the noise created by the equipment used to amplify the signal, and indeed it may be feebler. Under such circumstances it would not be recognisable.

Retrieval of such a signal from the noise may be achieved by averaging the output from the receiver over a time of, say, several seconds.

9.7   DEMONSTRATION 7

Retrieval of a radio signal demonstration

The appearance on a cathode ray tube or by a pen recorder of the noise generated by a receiver, is very much like grass whose height varies in a random way along the length of the time base. For simplicity in this demonstration the time base, taken as 200 mm long, is divided into 20 widths each 10 mm wide. The receiver noise in each width has a height which is proportional to a number from 0 to 10 generated at random by a computer and representing the strength of noise at that time in units of 10 mm. The signal in this case has a constant strength represented by 50 mm in the 15th width, but, of course, it could have any height in any place.

Histograms of signals showing for each time taken are drawn with a felt-tipped marker pen on separate overhead transparency films. The marker pens should produce coloured areas which are semi-transparent and of as uniform density as possible.

Looking at the histograms separately, one cannot say where the signal is, but if the transparencies are laid on top of one another, the signal and its position become easily recognisable.

Figure 9.9 shows a typical histogram (that for Time A in Table 9.1).

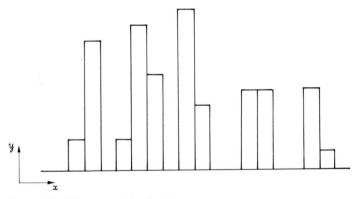

Figure 9.9   Histogram of signal noise

The output is normally recorded by a trace from a pen recorder, but it will be seen that the same method will apply if one imagines the trace to be divided up into strips at right angles to the time axis.

TABLE 9.1 Noise strengths and positions

Time A		Time B		Time C		Time D		Time E	
x	y	x	y	x	y	x	y	x	y
11	4	20	6	14	5	18	5	10	8
8	6	9	8	5	9	1	10	6	5
14	5	18	2	18	10	4	2	5	9
3	2	3	7	3	2	2	9	4	10
4	8	14	9	10	7	10	7	19	1
18	5	2	10	6	8	9	5	3	1
7	9	7	3	7	5	3	2	11	4
10	10	6	1	8	9	14	6	8	7
19	1	11	3	17	6	20	5	14	1
6	2	12	3	19	7	6	8	16	1
15	5	15	5	15	5	15	5	15	5

Time F		Time G		Time H		Time J		Time K	
x	y	x	y	x	y	x	y	x	y
3	7	13	10	20	8	8	1	8	2
14	8	20	5	12	2	10	1	17	7
1	8	10	10	13	8	3	5	19	10
4	5	3	3	6	2	13	2	14	4
18	2	8	1	8	8	20	7	4	5
6	6	14	4	14	4	4	8	16	1
8	6	4	7	5	6	19	4	12	4
16	3	11	9	3	6	9	7	13	8
20	2	16	6	2	4	14	6	18	3
5	4	17	7	9	8	18	6	1	8
15	5	15	5	15	5	15	5	15	5

Where a distance x along the time base is omitted, for example at $x = 5$ in Time A, the signal strength at that position is zero.

Appendices

A1 NOMOGRAPH MATHEMATICS (Section 2.17)

The nomographs referred to here will be restricted to those relating three variable quantities only. For a fuller discussion the reader is referred to *The Nomogram*, by Allcock, Jones and Michel, published by Pitman and Sons Ltd., or to *Graphic Aids in Engineering Computation*, by Hoelscher, Arnold and Pierce, published by McGraw Hill Company, Inc.

The equation relating the quantities to be represented must first be put in determinant form and the rows and columns manipulated according to the rules for determinants so that all the terms in the last column are unity.

The determinant will then have the form

$$\begin{vmatrix} g_1(u) & f_1(u) & 1 \\ g_2(v) & f_2(v) & 1 \\ g_3(w) & f_3(w) & 1 \end{vmatrix} = 0$$

where $g_1(u)$, $f_1(u)$ . . . etc., are functions of the variables u, v, w. It is important to note that each row must contain one of the variables only.

The loci of u, v and w may be drawn by plotting on x, y axes the positions of the points $x_1 = g_1(u)$, $y_1 = f_1(u)$ for selected values of u; $x_2 = g_2(v)$, $y = f_2(v)$ for selected values of v and $x_3 = g_3(w)$, $y_3 = f_3(w)$ for selected values of w.

A2 NOMOGRAPH RELATING MASS WITH SIZE AND PERIOD OF ORBIT (Project 8)

For the nomograph relating mass, size of orbit and period of orbit (Project 8) the equation relating the variables M, a and P is

$$P^2 = \frac{4\pi^2}{k^2 M} a^3$$

Taking logarithms to base 10 of both sides we have

$$2 \log_{10} P = 3 \log_{10} a - \log_{10} M + \log \frac{4\pi^2}{k^2}$$

or,

$$2 \log_{10} P - 2 \log_{10} \left(\frac{2\pi}{k} \right) - 3 \log_{10} a + \log_{10} M = 0$$

In determinant form this is

$$\begin{vmatrix} \log_{10} M & 0 & 1 \\ -3 \log_{10} a & 1 & 0 \\ -2 \log_{10} P + 2 \log_{10} \frac{2\pi}{k} & 1 & 1 \end{vmatrix} = 0$$

Divide the second row by -3 and the third row by -2

$$\begin{vmatrix} \log M & 0 & 1 \\ \log a & -\frac{1}{3} & 0 \\ \log P - \log \frac{2\pi}{k} & -\frac{1}{2} & -\frac{1}{2} \end{vmatrix} = 0$$

Add the second column to the third column

$$\begin{vmatrix} \log M & 0 & 1 \\ \log a & -\frac{1}{3} & -\frac{1}{3} \\ \log P - \log \frac{2\pi}{k} & -\frac{1}{2} & -1 \end{vmatrix} = 0$$

Multiply the second column by -1

$$\begin{vmatrix} \log M & 0 & 1 \\ \log a & \frac{1}{3} & -\frac{1}{3} \\ \log P - \log \frac{2\pi}{k} & -\frac{1}{2} & -1 \end{vmatrix} = 0$$

Multiply the second column by -4 and add to the third column

$$\begin{vmatrix} \log M & 0 & 1 \\ \log a & \frac{1}{3} & 1 \\ \log P - \log \frac{2\pi}{k} & \frac{1}{2} & 1 \end{vmatrix} = 0$$

To achieve vertical scales interchange the second and third columns

$$\begin{vmatrix} 0 & \log M & 1 \\ \dfrac{1}{3} & \log a & 1 \\ \dfrac{1}{2} & \log P - \log \dfrac{2\pi}{k} & 1 \end{vmatrix} = 0$$

Finally multiply the first column by 6

$$(x)\ (y)$$

$$\begin{vmatrix} 0 & \log M & 1 \\ 2 & \log a & 1 \\ 3 & \log P - \log \dfrac{2\pi}{k} & 1 \end{vmatrix} = 0$$

Thus since all the x co-ordinates are constant the M line will be the line $y = 0$; the a line will be the line $y = 2$ and the P line will be the line $y = 3$ to any convenient scale we care to choose.

The M line can be calibrated from $y = \log_{10} M$ by giving values of M and calculating the corresponding y. Similarly for the a line, and similarly for the P line, $y = \log P - \log 2\pi/k$ where the scale will be depressed by $\log_{10} 2\pi/k$ where $k = 0.0172$.

A3 (a) NOMOGRAPH RELATING HOUR ANGLE H, ZENITH DISTANCE z AND DELINATION δ (Project 14)

For the nomograph relating the hour angle H, zenith distance z and declination δ the equation relating the variables H, z and δ is

$$\cos z = \sin \varphi \sin \delta + \cos \varphi \cos \delta \cos H$$

or,
$$\cos z - \sin \varphi \sin \delta - \cos \varphi \cos \delta \cos H = 0$$

from the spherical triangle PZX.

In determinant form this becomes

$$\begin{vmatrix} -\cos \varphi \cos H & 0 & 1 \\ \cos z & 1 & 0 \\ \sin \varphi \sin \delta & 1 & \cos \delta \end{vmatrix} = 0$$

Adding the second column to the third

$$\begin{vmatrix} -\cos\varphi\cos H & 0 & 1 \\ \cos z & 1 & 1 \\ \sin\varphi\sin\delta & 1 & 1+\cos\delta \end{vmatrix} = 0$$

Dividing the third row by $(1+\cos\delta)$

$$\begin{vmatrix} -\cos\varphi\cos H & 0 & 1 \\ \cos z & 1 & 1 \\ \sin\varphi.\dfrac{\sin\delta}{1+\cos\delta} & \dfrac{1}{1+\cos\delta} & 1 \end{vmatrix} = 0$$

$\dfrac{\sin\delta}{1+\cos\delta}$ reduces to $\dfrac{2\,\sin\delta/2\,\cos\delta/2}{1+2\,\cos^2\delta/2-1} = \tan\delta/2$

Substituting this and at the same time interchanging the first and second columns

$$\begin{matrix} (x) & (y) & \\ \begin{vmatrix} 0 & -\cos\varphi\cos H & 1 \\ 1 & \cos z & 1 \\ \dfrac{1}{1+\cos\delta} & \sin\varphi\tan\delta/2 & 1 \end{vmatrix} = 0 \end{matrix}$$

The x, y co-ordinates in each row give the scale for each variable

(b) NOMOGRAPH RELATING AZIMUTH A, ZENITH DISTANCE z AND DECLINATION δ (Project 14)

For the nomograph relating the azimuth A, zenith distance z and declination δ the equation is, from spherical triangle PZX

$$\sin\delta = \sin\varphi\cos z + \cos\varphi\sin z\cos A$$

or, $\cos\varphi\cos A\sin z + \sin\varphi\cos z - \sin\delta = 0.$

In determinant form this becomes

$$\begin{vmatrix} -\cos\varphi\cos H & 0 & 1 \\ \cos z & 1 & 0 \\ \sin\varphi\sin\delta & 1 & \cos\delta \end{vmatrix} = 0$$

Following the same procedure as before

$$\begin{vmatrix} (x) & (y) & \\ 0 & -\cos\varphi\cos A & 1 \\ 1 & \sin\delta & 1 \\ \dfrac{1}{1+\sin z} & \dfrac{\sin\varphi\cos z}{1+\sin z} & 1 \end{vmatrix} = 0$$

The x, y co-ordinates in each row now give the scale for each variable,

$\dfrac{\cos z}{1+\sin z}$ reduces to $\dfrac{1-\tan\delta/2}{1+\tan\delta/2}$ if this is preferred.

A4   RELATION BETWEEN φ, δ AND a AT RISING AND SETTING (Project 32)

Referring to spherical triangle NPX (Figure 5.21) from the cosine formula

$$\sin\delta = \sin a \cos\varphi \quad \text{since} \quad \angle N = 90°$$

or, $\sin a = \sec\varphi \sin\delta$

A5   RISING AND SETTING OF THE SUN FORMULA (Project 32)

The equation relevant to the rising or setting of the Sun is (See A4)

$$\sin a = \sec\varphi \sin\delta$$

or, $\sin a - \sec\varphi \sin\delta = 0$

Putting this in determinant form

$$\begin{vmatrix} \sin a & 0 & 1 \\ \sec\varphi & 1 & 0 \\ 0 & -\sin\delta & 1 \end{vmatrix} = 0$$

Add the third column to the first column

$$\begin{vmatrix} 1+\sin a & 0 & 1 \\ \sec\varphi & 1 & 0 \\ 1 & -\sin\delta & 1 \end{vmatrix} = 0$$

Divide the first row by $(1+\sin a)$ and the second row by $\sec \varphi$

$$\begin{vmatrix} 1 & 0 & \dfrac{1}{1+\sin a} \\ 1 & \cos \varphi & 0 \\ 1 & -\sin \delta & 1 \end{vmatrix} = 0$$

Interchange the first and third columns

$$\begin{matrix} (x) & & (y) \\ \end{matrix}$$
$$\begin{vmatrix} \dfrac{1}{1+\sin a} & 0 & 1 \\ 0 & \cos \varphi & 1 \\ 1 & -\sin \delta & 1 \end{vmatrix} = 0$$

The x, y co-ordinates give the nomograph curve for each variable a, φ and δ and should be plotted to a scale of 10 mm = 0.1 unit.

A6 NOMOGRAPH RELATING APPARENT MAGNITUDE m, ABSOLUTE MAGNITUDE M, AND PARALLAX p (Project 41)

The equation connecting the apparent magnitude m, the absolute magnitude M and the parallax p of a star is

$$M = m+5+5 \log_{10} p$$

or,
$$M-(m+5)-5 \log_{10} p = 0$$

In determinant form this becomes

$$\begin{vmatrix} M & 1 & 1 \\ (m+5) & 1 & 0 \\ 5 \log_{10} p & 0 & 1 \end{vmatrix} = 0$$

Add the second column to the third column and then divide the resulting first row by 2

$$\begin{vmatrix} \tfrac{1}{2}M & \tfrac{1}{2} & 1 \\ (m+5) & 1 & 1 \\ 5 \log_{10} p & 0 & 1 \end{vmatrix} = 0$$

Interchange the first and second rows

$$\begin{vmatrix} (m+5) & 1 & 1 \\ \frac{1}{2}M & \frac{1}{2} & 1 \\ 5 \log_{10} p & 0 & 1 \end{vmatrix} = 0$$

Interchange the first and second columns

$$\begin{vmatrix} 1 & (m+5) & 1 \\ \frac{1}{2} & \frac{1}{2}M & 1 \\ 0 & 5 \log_{10} p & 1 \end{vmatrix} = 0$$

Thus the graphs to be drawn are

$$x = 1, \quad y = m+5$$
$$x = \tfrac{1}{2}, \quad y = \tfrac{1}{2}M$$
$$x = 0, \quad y = 5 \log_{10} p$$

A7 LOCATION OF THE CENTRES O AND O'
(Project 27)

If we do not use a model, but can nevertheless determine the positions of O and O' in Figure 5.9 relative to the line YY', we can proceed with the construction on a flat board or card.

From Figure 5.9, if r is the length of the style OO',

then $MO = r/\cos \varphi$

and $MO' = r \tan \varphi$

Also, it is then possible to find the angles necessary on the sundial B between MO and the individual hour lines.

For instance, from Figure 5.9 the angle on B between MO and the second hour line $O2$ is given by

$\tan \theta_2 = M2/MO$ where $MO = r/\cos\varphi$ as above

Also on A $\tan 30° = M2/MO'$ where $MO' = r \tan \varphi$ as above

Thus $\tan \theta_2 = \dfrac{MO' \tan 30°}{r/\cos\varphi} = \dfrac{r \tan \varphi \tan 30° \cos \varphi}{r}$

or, $\tan \theta_2 = \sin \varphi \tan 30°$

Similarly $\tan \theta_3 = \sin \varphi \tan 45°$ and so on.

For latitude $\varphi = 52°$ the angles on the sundial from MO and the hour lines from 0 to 6 hours are respectively

noon to 1 hour $= 11°55'$
noon to 2 hours $= 24°28'$
noon to 3 hours $= 38°15'$
noon to 4 hours $= 53°47'$
noon to 5 hours $= 71°13'$
noon to 6 hours $= 90°$

All the other lines can be obtained either by symmetry or by extension of these lines.

A8   FLOW OF PHOTONS FROM THE SUN (Section 5.24)

The mathematics involved in the demonstrations illustrating the flow of photons from the Sun is based on a probability topic known as the 'random walk'. Although Demonstration 4 is two-dimensional, the three-dimensional case follows the same theory and, perhaps unexpectedly, gives the same result, as indeed does the one-dimensional case.

As an introduction, the one-dimensional case will be considered, in which the photon is restricted to absorptions and re-emissions in the same straight line. The distance travelled between emission and re-absorption, the mean free path, will be designated a step s. The probability of forward emission $(+s)$ is the same as backward emission $(-s)$, so that we would expect that the mean of the distances (x_n) from the central position after a large number of steps n would be zero, that is $\Sigma x_n/n = 0$.

However, it is clear that after a given number of steps (n) the photon will be at some distance (x_n) from the central position, and statistically the standard deviation (σ) will give this value, where $\sigma^2_n = \Sigma(x^2_n)/n$.

Referring to Figure A1 in which O is the central position, P the position of the photon after $(n-1)$ steps and P' the position after one further step s, that is, the position after n steps,

then
$$x_n = x_{n-1} \pm s$$

$$\sigma^2_n = \frac{\Sigma x^2_n}{n} = \frac{\Sigma x^2_n}{n} - 1 \pm \frac{2s\Sigma x_{n-1}}{n} + \frac{ns^2}{n}$$

Since
$$\frac{\Sigma x_{n-1}}{n} = 0, \ \sigma^2_n = \sigma^2_{n-1} + s^2$$

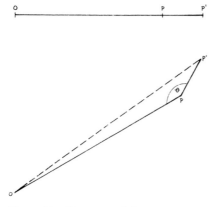

Figure A1   Movement of photon

This is now a reduction formula, and we can obtain σ_n in terms of σ_1.

$$\sigma_{n-1}^2 = \sigma_{n-2}^2 + s^2$$

giving

$$\sigma_n^2 = \sigma_{n-2}^2 + s^2 + s^2 = \sigma_{n-2}^2 + 2s^2$$

and

$$\sigma_n^2 = \sigma_1^2 + (n-1)\,s^2$$

But

$$\sigma_1^2 = \frac{s^2}{1}$$

so

$$\sigma_n^2 = s^2 + (n-1)\,s^2$$

or

$$\underline{\sigma_n^2 = n\,s^2}$$

For the three dimensional case, referring to Figure A1 and using the cosine rule

$$x_n^2 = x_{n-1}^2 + s^2 - 2x_{n-1}.s.\cos\theta$$

where θ defines the direction of the nth step which can be in the plane of the paper, but more generally out of the plane of the paper.

If we now recognise that the term

$$\frac{2s\,\Sigma x_{n-1}.\cos\theta}{n} = 0,$$

the analysis follows that for the one-dimensional case closely, with the same result, namely

$$\sigma_n^2 = ns^2$$

For the case of the Sun the steps are about 10 mm, the mean free path, and $\sigma_n = 7 \times 10^{11}$ mm, the radius of the Sun. Thus the number of steps n required for the photon to reach the surface of the Sun is given by

$$(7 \times 10^{11})^2 = n.10^2$$

or $$n = 5 \times 10^{21} \quad \text{approximately}$$

The speed of light is 3×10^{11} mm/s so that the time taken for a photon to travel a distance of $5 \times 10^{21} \times 10$ mm is about 2×10^{11} seconds or about 6000 years.

A9   NOCTURNAL MATHEMATICS (Project 34)

The correct relative setting of the time circle and the date circle for the nocturnal may be deduced as follows,

When Dubhe $(a = 11$ h, $\delta = 62°)$ is in the north (azimuth $0°$) and above Polaris, its hour angle H is $0°$.

But since $a + H = $ local sidereal time (see Section 4.4),
local sidereal time $= 11$ h $+ 0 = 11$ h.

From tables such as Time Reckoning in the *BAA Handbook* we find the date to be very nearly midnight on March 7. Thus the radius through March 7 should also pass through one of the 12 hour markings.

A10   ASTROLABE MATHEMATICS (Project 35)

The projection for the plate markings of the astrolabe is south polar stereographic onto the equatorial plane.

Referring to Figure A2, imagine a plane through the equator. This is to be the plane of projection. Draw any star A on the celestial sphere. The line joining A to S, the south celestial pole, will cut the equatorial plane in, say, A_1. The map of the heavens is made up of many similarly constructed points.

Consider a basic sphere, to represent the celestial sphere, of radius r, say. Axes can be set up with the south pole S as the origin so that the north pole N and the observer's zenith Z lie in the plane zSx, Figure A3.

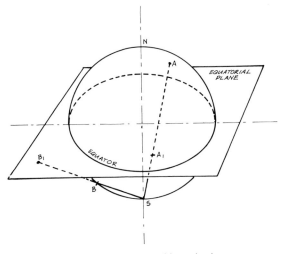

Figure A2   South polar stereographic projection

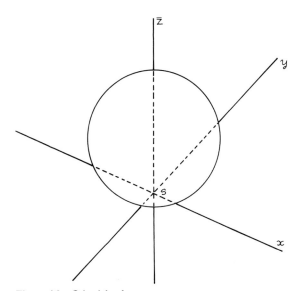

Figure A3   Celestial sphere axes

The equation of this sphere is

$$x^2+y^2+(z-r)^2 = r^2$$

or,

$$x^2+y^2+z^2-2rz = 0 \qquad [1]$$

Let *any* plane cutting the sphere have equation

$$lx + my + nz = p$$

so that

$$(lx + my + nz)/p = 1 \qquad [2]$$

Consider now the equation

$$x^2 + y^2 + z^2 - 2rz(lx + my + nz)/p = 0 \qquad [3]$$

This is a homogeneous equation in x, y, z and therefore represents *a* cone with its vertex at the origin. But the co-ordinates of sphere and plane satisfy [1] and [2] respectively and therefore satisfy [3].

Hence equation [3] represents *the* cone, vertex at the origin S, through the intersection of the sphere [1] and the plane [2]. This intersection will be a general circle on the celestial sphere.

The cone cuts the equatorial plane $z = r$ where, from [3],

$$x^2 + y^2 + r^2 - 2r^2(lx + my + nr)/p = 0 \qquad [4]$$

Equation [4] is that of a circle on the equatorial plane. Thus the polar stereographic projection of *any* circle on to the equatorial plane is another circle.

This is a most useful property of polar stereographic projection since we shall be concerned with the projection of circles of equal altitude (small circles) and circles of equal azimuth (great circles); with circles of declination (small circles) and circles of right ascension (great circles).

Looking now at the sphere cut by the plane zSx, Figure A4, NS is the

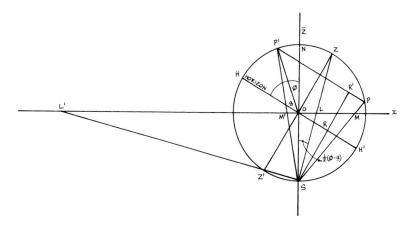

Figure A4

north–south line; the horizon of the observer O makes an angle φ with NS since the altitude of the north pole is the latitude λ of the observer. Let P, P' be points of equal altitude a and Z the zenith.

The direction cosines of the perpendicular to the plane through PP' parallel to the plane of the horizon are $\cos \varphi, 0$ and $\sin \varphi$.

Also the length of this perpendicular p is given by SR'

$$p = SR + RR' = r \sin \varphi + r \sin a$$

Thus the equation of this plane is

$$\cos \varphi . x + \sin \varphi . z = r \sin \varphi + r \sin a$$

or, $(\cos \varphi . x + \sin \varphi . z)/r(\sin \varphi + \sin a) = 1$

Using [4] the image of the small circle through PP' parallel to the horizon is therefore

$$x^2 + y^2 + r^2 - 2r(\cos \varphi . x + \sin \varphi . r)/(\sin \varphi + \sin a) = 0$$

or, $x^2 + y^2 - 2r \cos \varphi . x/(\sin \varphi + \sin a) - r^2(\sin \varphi - \sin a)/(\sin \varphi + \sin a) = 0$

Thus, for the circle of constant elevation a on the plate the centre has co-ordinates

$$x = r \cos \varphi/(\sin \varphi + \sin a), \quad y = 0 \qquad [5]$$

The radius of this circle is

$$\{r^2 \cos^2 \varphi/(\sin \varphi + \sin a)^2 + r^2(\sin \varphi - \sin a)/(\sin \varphi + \sin a)\}^{\frac{1}{2}}$$

which reduces to

$$\text{radius} = r \cos a/(\sin \varphi + \sin a) \qquad [6]$$

It is of mathematical interest to note that all the circles of constant azimuth pass through the projections of Z, Z' on to the plate. They thus form a set of co-axal circles with the x-axis as radical axis.

Similarly, the equation of a typical circle of constant elevation is, as before

$$x^2 + y^2 - 2r \cos \varphi . x/(\sin \varphi + \sin a) - r^2(\sin \varphi - \sin a)/(\sin \varphi + \sin a) = 0$$

For any two such circles of elevation a and β respectively, the radical axis will be given by

$$g.x + f.y + c = g'.x + f'.y + c'$$

(usual circle notation).

The radical axis is therefore

$$r \cos \varphi . x/(\sin \varphi + \sin a) - r^2(\sin \varphi - \sin a)/(\sin \varphi + \sin a)$$
$$= r \cos \varphi . x/(\sin \varphi + \sin \beta) - r^2(\sin \varphi - \sin \beta)/(\sin \varphi + \sin \beta)$$

which reduces to

$$x = -r \tan \varphi$$

Thus it will be seen that the circles of elevation form a co-axal system with radical axis $x = -r \tan \varphi$ and also that the two co-axal systems are orthogonal.

Consider next the circle of constant bearing $\bar{a}$ measured from the north and which will pass through Z, Z' and make an angle $\bar{a}$ with the zOx plane.

The direction cosines of the perpendicular to the plane of constant bearing $\bar{a}$ are (Figure A5):

$$l = \sin \bar{a} \sin \varphi$$
$$m = \cos \bar{a}$$
$$n = -\sin \bar{a} \cos \varphi$$

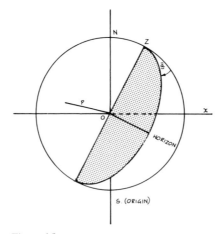

Figure A5

Thus the equation of this plane of constant bearing or azimuth $\bar{a}$ is

$$\sin \bar{a} \sin \varphi . x + \cos \bar{a} . y - \sin \bar{a} \cos \varphi . z = p$$

But the plane passes through the point $(0, 0\ r)$, so that

$$p = -r \sin \bar{a} \cos \varphi$$

and the plane is

$$-(\sin \bar{a} \sin \varphi.x+\cos \bar{a}.y-\sin \bar{a} \cos \varphi.z)/r \sin \bar{a} \cos \varphi = 1$$

Thus the equation of the cone with S as vertex and cutting the sphere in the circle ZHZ' is

$$x^2+y^2+z^2+2zr(\sin \bar{a} \sin \varphi.x+\cos \bar{a}.y-\sin \bar{a} \cos \varphi.z)/r \sin \bar{a} \cos \varphi = 0$$

This cuts the equatorial plane $z = r$ in the circle

$$x^2+y^2+2r \tan \varphi.x+2r \cot \bar{a} \sec \varphi.y-2r^2 = 0$$

Hence the circle on the plate with constant azimuth $\bar{a}$ has its centre at co-ordinates

$$(x = -r \tan \varphi, \quad y = -r \cot \bar{a} \sec \varphi)$$

The radius of this circle can be calculated but this is of academic interest since we know that all the circles of constant azimuth pass through the projection of the zenith Z and the nadir Z'. Also, since φ is constant for any one plate of the astrolabe, these centres all lie on $x = -r \tan \varphi$, a line parallel to the y axis.

A11   RELATION BETWEEN THE RADIUS OF THE BASE SPHERE AND THAT OF THE PLATE (Project 35)

The outer limit of the plate circle is taken to be that of the projection of the Tropic of Capricorn onto the equatorial plane. This is the most southerly point which the Sun reaches in its passage along the ecliptic, Figure A6.

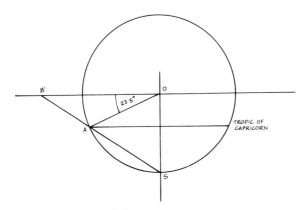

Figure A6   Radius of plate

Let R, r = radius of plate circle, radius of base sphere.

Then   angle OSA = 56.75°

and $OB'/OS = R/r = \tan 56.75° = 1.53$

or, $R = 1.53 r$

A12   RELATION BETWEEN THE HOUR ANGLE (H.A.) OF A STAR X AND THE RIGHT ASCENSION (R.A.) OF THE STAR X (Project 35)

The relation is shown in standard texts as:

H.A. of X + R.A. of X = H.A. of the First Point of Aries.

But, by lining up the First Point of Aries on the rete with zero hour angle, we have automatically pre-set the astrolabe so that

H.A. of X + R.A. of X = 0°   or 360°.

Thus the hour angle of X given in the table = 360° − right ascension of X.

Normally, in quoting the co-ordinates of a star, right ascension is given and this must be subtracted from 360° to obtain the 'astrolabe' hour angle.

A13   LINES OF CONSTANT ALTITUDE AND AZIMUTH (Project 35)

For r = 50 mm   and φ = 52°, r_1 = 32.8 mm, r_2 = 76.3 mm   and x_A = − 64.0 mm.

TABLE A1

a (°)	0	10	20	30	40	50	60	70	80	90
R(mm)	63.4	51.2	41.6	33.6	26.7	20.4	15.1	9.9	4.9	0
x_a(mm)	39.1	31.8	27.2	23.9	21.5	19.8	18.6	17.8	17.3	17.2

A(°)	0	10	20	30	40	50	60	70	80	90
y_A(mm)	− ∞	−460	−223	−140	−96.6	−68.2	−47.1	−29.2	−14.6	0

A(°)	100	110	120	130	140	150	160	170	180
y_A(mm)	14.6	29.2	47.1	68.2	96.6	140	223	460	∞

A14   LINEAR ASTROLABE TABLES (Project 36)

The markings on the plate have been obtained as (x, y) coordinates and are tabulated below, assuming a latitude of 52 °N.

TABLE A2

Altitude									
0°		10°		20°		30°		40°	
x	y	x	y	x	y	x	y	x	y
0	5.9	0	7.6	0	9.5	0	11.5	0	13.8
3.3	5.6	3.8	7.3	4.5	9.1	5.7	11.0	8.1	12.9
6.3	4.9	7.2	6.5	8.3	8.0	9.9	9.6	12.5	11.1
9.0	3.9	10.0	5.3	11.3	6.7	12.9	8.0	15.1	9.3
11.4	2.7	12.5	4.0	13.8	5.2	15.3	6.4	17.1	7.6
13.6	1.4	14.7	2.6	15.9	3.8	17.3	4.9	18.8	6.0
15.7	0	16.8	1.2	17.9	2.3	19.1	3.5	20.5	4.6
17.8	−1.4	18.8	−0.2	19.9	1.0	21.0	2.2	22.1	3.4
20.0	−2.7	21.0	−1.4	21.9	−0.2	22.8	1.1	23.8	2.4
22.4	−3.9	23.3	−2.6	24.1	−1.2	24.8	0.2	25.6	1.5
25.1	−4.9	25.8	−3.4	26.4	−2.0	26.9	−0.6	27.5	0.8
28.1	−5.6	28.5	−4.0	28.8	−2.5	29.1	−1.0	29.4	0.4
31.4	−5.9	31.4	−4.3	31.4	−2.7	31.4	−1.2	31.4	0.3
34.7	−5.6	34.3	−4.0	34.0	−2.5	33.7	−1.0	33.4	0.4
37.7	−4.9	37.0	−3.4	36.5	−2.0	35.9	−0.6	35.4	0.8
40.4	−3.9	39.6	−2.6	38.8	−1.2	38.0	0.2	37.2	1.5
42.8	−2.7	41.9	−1.4	40.9	−0.2	40.0	1.1	39.0	2.4
45.0	−1.4	44.0	−0.2	43.0	1.0	41.9	2.2	40.7	3.4
47.1	0	46.0	1.2	44.9	2.3	43.7	3.5	42.4	4.6
49.2	1.4	48.1	2.6	46.9	3.8	45.6	4.9	44.0	6.0
51.4	2.7	50.3	4.0	49.1	5.2	47.6	6.4	45.7	7.6
53.8	3.9	52.8	5.3	51.5	6.7	49.9	8.0	47.7	9.3
56.5	4.9	55.7	6.5	54.5	8.0	52.9	9.6	50.3	11.1
59.6	5.6	59.1	7.3	58.3	9.1	57.2	11.0	54.7	12.9
62.8	5.9	62.8	7.6	62.8	9.5	62.8	11.5	62.8	13.8

Altitude

50°		60°		70°		80°		90°	
x	y	x	y	x	y	x	y	x	y
0	16.5	31.4	14.8	31.4	12.4	31.4	10.3	31.4	8.3
14.6	14.4	24.4	13.9	28.7	12.1	30.5	10.2		
16.7	12.2	22.3	12.3	26.8	11.4	29.6	9.9		
18.2	10.3	22.0	10.7	25.8	10.4	29.0	9.5		
19.4	8.6	22.3	9.2	25.5	9.4	28.7	9.1		
20.7	7.1	23.0	7.9	25.7	8.4	28.6	8.6		
22.0	5.7	23.9	6.7	26.1	7.6	28.6	8.1		
23.4	4.6	24.9	5.7	26.7	6.8	28.9	7.7		
24.9	3.6	26.1	4.9	27.5	6.1	29.2	7.3		
26.4	2.8	27.3	4.2	28.4	5.6	29.7	7.0		
28.0	2.3	28.6	3.7	29.3	5.2	30.2	6.7		
29.7	1.9	30.0	3.4	30.4	5.0	30.8	6.6		
31.4	1.8	31.4	3.3	31.4	4.9	31.4	6.6		
33.1	1.9	32.8	3.4	32.5	5.0	32.0	6.6		
34.8	2.3	34.2	3.7	33.5	5.2	32.6	6.7		
36.4	2.8	35.5	4.2	34.5	5.6	33.1	7.0		
38.0	3.6	36.8	4.9	35.4	6.1	33.6	7.3		
39.4	4.6	37.9	5.7	36.1	6.8	34.0	7.7		
40.8	5.7	39.0	6.7	36.8	7.6	34.2	8.1		
42.1	7.1	39.8	7.9	37.2	8.4	34.3	8.6		
43.4	8.6	40.5	9.2	37.3	9.4	34.1	9.1		
44.7	10.3	40.9	10.7	37.0	10.4	33.8	9.5		
46.1	12.2	40.6	12.3	36.0	11.4	33.2	9.9		
48.2	14.4	38.4	13.9	34.1	12.1	32.4	10.2		
62.8	16.5	31.4	14.8	31.4	12.4	31.4	10.3		

TABLE A3

Azimuth

0°		10°		20°		30°		40°	
x	y	x	y	x	y	x	y	x	y
31.4	8.3	31.4	8.3	31.4	8.3	31.4	8.3	31.4	8.3
	5.7	32.0	5.7	32.5	5.8	33.1	6.0	33.6	6.2
	3.3	32.4	3.4	33.3	3.5	34.2	3.7	35.1	4.0
	1.0	32.7	1.1	33.9	1.3	35.1	1.6	36.3	1.9
	−1.2	32.9	−1.1	34.4	−0.9	35.9	−0.6	37.3	−0.1
	−3.5	33.2	−3.4	35.0	−3.1	36.8	−2.7	38.4	−2.2
	−5.9	33.6	−5.8	35.7	−5.4	37.7	−4.9	39.6	−4.3
	−8.5	34.2	−8.4	36.8	−7.9	39.1	−7.2	41.1	−6.4
	−11.5	35.3	−11.2	38.7	−10.6	41.3	−9.6	43.4	−8.6
	−15.1	39.7	−14.4	43.6	−13.1	45.8	−11.8	47.3	−10.4
	−14.8	57.4	−14.4	54.7	−13.4	53.7	−12.3	53.4	−11.2
	−11.3	61.7	−11.2	60.7	−11.0	59.8	−10.7	59.2	−10.3
	−8.3	62.8	−8.3	62.8	−8.3	62.8	−8.3	62.8	−8.3
	−5.7	0.6	−5.7	1.1	−5.8	1.7	−6.0	2.2	−6.2
	−3.3	0.9	−3.4	1.9	−3.5	2.8	−3.7	3.7	−4.0
	−1.0	1.2	−1.1	2.5	−1.3	3.7	−1.6	4.8	−1.9
	1.2	1.5	1.1	3.0	0.9	4.5	0.6	5.9	0.1
	3.5	1.8	3.4	3.6	3.1	5.3	2.7	7.0	2.2
	5.9	2.2	5.8	4.3	5.4	6.3	4.9	8.2	4.3
	8.5	2.8	8.4	5.4	7.9	7.7	7.2	9.7	6.4
	11.5	3.9	11.2	7.3	10.6	9.9	9.6	12.0	8.6
	15.1	8.2	14.4	12.2	13.1	14.4	11.8	15.9	10.4
	14.8	26.0	14.4	23.3	13.4	22.3	12.3	22.0	11.2
	11.3	30.3	11.2	29.2	11.0	28.4	10.7	27.8	10.3
31.4	8.3	31.4	8.3	31.4	8.3	31.4	8.3	31.4	8.3

Azimuth

50°		60°		80°		100°		120°	
x	y	x	y	x	y	x	y	x	y
31.4	8.3	31.4	8.3	31.4	8.3	31.4	8.3	31.4	8.3
34.1	6.4	34.5	6.7	35.3	7.4	35.7	8.3	35.7	9.3
36.0	4.4	36.8	4.9	38.3	6.0	39.6	7.5	40.5	9.2
37.4	2.4	38.5	3.0	40.6	4.4	42.6	6.1	44.6	8.1
38.7	0.4	40.0	1.1	42.5	2.6	44.9	4.4	47.6	6.4
39.9	−1.5	41.4	−0.8	44.2	0.9	46.8	2.7	49.7	4.6
41.3	−3.5	42.8	−2.7	45.7	−0.9	48.5	0.9	51.4	2.7
42.9	−5.5	44.5	−4.6	47.4	−2.7	50.1	−0.9	52.9	0.8
45.2	−7.5	46.7	−6.4	49.3	−4.4	51.8	−2.6	54.3	−1.1
48.5	−9.2	49.6	−8.1	51.6	−6.1	53.7	−4.4	55.7	−3.0
53.4	−10.2	53.7	−9.2	54.7	−7.5	56.0	−6.0	57.5	−4.9
58.8	−9.8	58.6	−9.3	58.6	−8.3	59.0	−7.4	59.7	−6.7
62.8	−8.3	62.8	−8.3	62.8	−8.3	62.8	−8.3	62.8	−8.3
2.7	−6.4	3.1	−6.7	3.8	−7.4	4.3	−8.3	4.3	−9.3
4.5	−4.4	5.4	−4.9	6.9	−6.0	8.1	−7.5	9.1	−9.2
6.0	−2.4	7.1	−2.9	9.2	−4.4	11.2	−6.1	13.2	−8.1
7.3	−0.4	8.6	−1.1	11.1	−2.6	13.5	−4.4	16.2	−6.4
8.5	1.5	10.0	0.8	12.7	−0.9	15.4	−2.7	18.3	−4.6
9.9	3.5	11.4	2.7	14.3	0.9	17.1	−0.9	20.0	−2.7
11.5	5.5	13.1	4.6	16.0	2.7	18.7	0.9	21.4	−0.8
13.8	7.5	15.3	6.4	17.9	4.4	20.8	2.6	22.8	1.1
17.1	9.2	18.2	8.1	20.2	6.1	22.2	4.4	24.3	3.0
22.0	10.2	22.3	9.2	23.3	7.5	24.6	6.0	26.1	4.9
27.4	9.8	27.2	9.3	27.1	8.3	27.6	7.4	28.3	6.7
31.4	8.3	31.4	8.3	31.4	8.3	31.4	8.3	31.4	8.3

Azimuth

140°		160°		180°	
x	y	x	y	x	y
31.4	8.3	31.4	8.3	31.4	8.3
35.0	10.3	33.6	11.0	31.4	11.3
40.9	11.2	39.5	13.4	31.4	14.8
46.9	10.4	50.6	13.1	62.8	15.1
50.8	8.6	55.6	10.6	62.8	11.5
53.1	6.4	57.5	7.9	62.8	8.5
54.7	4.3	58.5	5.4	62.8	5.9
55.9	2.2	59.2	3.1	62.8	3.5
56.9	0.1	59.8	0.9	62.8	1.2
58.0	−1.9	60.4	−1.3	62.8	−1.0
59.2	−4.0	61.0	−3.5	62.8	−3.3
60.6	−6.2	61.7	−5.8	62.8	−5.7
62.8	−8.3	62.8	−8.3	62.8	−8.3
3.6	−10.3	2.2	−11.0	0	−11.3
9.5	−11.2	8.1	−13.4	0	−14.8
15.5	−10.4	19.2	−13.1	31.4	−15.1
19.4	−8.6	24.2	−10.6	31.4	−11.5
21.7	−6.4	26.0	−7.9	31.4	−8.5
23.2	−4.3	27.1	−5.4	31.4	−5.9
24.4	−2.2	27.8	−3.1	31.4	−3.5
25.5	−0.1	28.4	−0.9	31.4	−1.2
26.6	1.9	29.0	1.3	31.4	1.0
27.7	4.0	29.6	3.5	31.4	3.3
29.2	6.2	30.3	5.8	31.4	5.7
31.4	8.3	31.4	8.3	31.4	8.3

A15   CALCULATION OF ANGLES FOR LOGARITHMIC DISC (Demonstration 5)

For simplicity only one quarter of each ring will be considered. In Figure A.7 let p_k = the proportion of white in the kth ring from the centre.

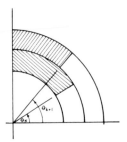

Figure A7   Logarithmic disc angles

If there are n rings, and the outer one is all white, then $p_n = 1$.

In general, $p_k = \dfrac{\theta_k}{90}$ if the angles are measured in degrees.

Again, if $\log p_{k+1} - \log p_k$ is constant for all values of k

then $\dfrac{p_{k+1}}{p_k} = c$, say, constant for all k.

For example, if there is to be an increase in white of say 10 per cent as we move from ring k to ring $k + 1$, then $c = 1.1$.

If we now multiply the $(n - 1)$ equal fractions

$$\frac{p_n}{p_{n-1}}, \quad \frac{p_{n-1}}{p_{n-2}}, \quad \ldots \quad \frac{p_2}{p_1}$$

we have $\dfrac{p_n}{p_1} = c^{n-1}$ since all other values cancel.

But $p_n = 1$, so $\dfrac{1}{p_1} = c^{n-1}$, and $p_1 = \dfrac{\theta_1}{90}$

Hence $\dfrac{90}{\theta_1} = c^{n-1}$

Finally, $\theta_1 c^{n-1} = 90$

Given two of the three quantities θ_1, c and n the third can be found.

For our case, $n = 6$ and if we choose a 43 per cent increase from ring to ring, then $c = 1.43$.

This then gives $\theta_1 = \dfrac{90}{1.43^5} = 15°$ approximately.

Also

$p_2 = c.p_1$ and $\theta_2 = 90.p_2 = 90.c.p_1 = 90.c.\theta_1/90 = c.\theta_1 = 21°$

Similarly

$$\theta_3 = c^2.\theta_1 = 31°; \quad \theta_4 = c^3.\theta_1 = 44°; \quad \theta_5 = c^4.\theta_1 = 63°$$

The value $c = 1.43$ was chosen so that the angles θ were both reasonably well spread over the quadrant and so that the smallest angle was measurable.

From an astronomy point of view it would be useful to arrange the value of c so that there would be exactly one magnitude difference between adjacent rings. This, however, gives very small angles for the three inner rings.

A compromise to this situation would be to settle for 0.5 magnitudes. The value of c is calculated from the basic magnitude formula

$$m_2 - m_1 = 2.5 \log_{10}(B_1/B_2)$$

Here $m_2 - m_1 = 0.5$, so that

$$\log_{10}(B_1/B_2) = 0.2$$

Hence

$$\log_{10}(p_{k+1}/p_k) = 0.2$$

or,

$$c = \frac{p_{k+1}}{p_k} = 1.585$$

The corresponding angles, calculated as described above, are

$$\theta_1 = 9°, \quad \theta_2 = 14°, \quad \theta_3 = 23°, \quad \theta_4 = 36° \quad \text{and} \quad \theta_5 = 57°$$

A16 DEVIATION OF RED LIGHT IN THE SPECTROMETER (Project 43)

From Figure 7.8 we see that

$$\sin \theta = \lambda/d \text{ where } \lambda = \text{wavelength of red light}$$
$$= 0.75 \times 10^{-3} \text{ mm, say}$$

d = distance between
diffraction grating lines
= 1/500 mm, say

Then $\sin \theta = 0.75 \times 10^{-3}/(1/500)$
$\sin \theta = 0.375$

or angle of deviation $\theta = 22°$ say

Thus the angle θ in Figure 7.12, and which controls the best shape of the box, should be about 22°.

Solutions

In the text all the projects have been designed to fit onto A4 size paper. The solutions in this book have been reduced in scale so that they can be accommodated on the page size of the book.

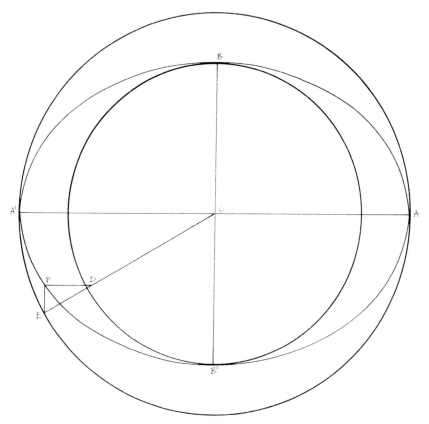

Project 1(i) Ellipse

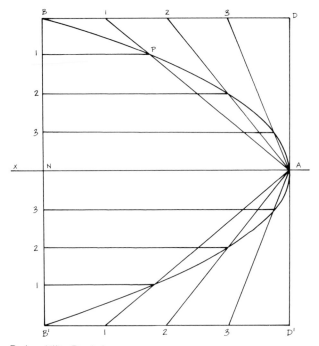

Project 1(ii)   Parabola

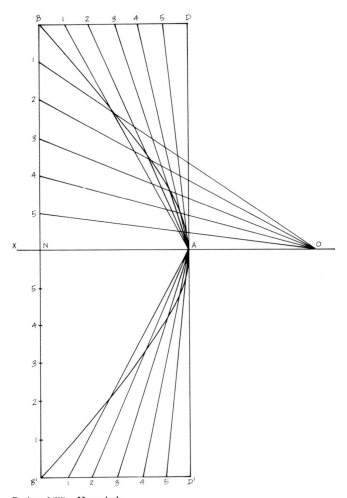

Project 1(iii) Hyperbola

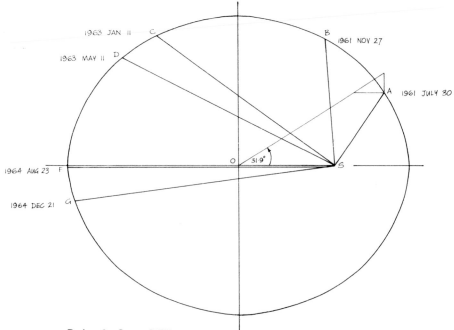

Project 2   Comet P/Wirtanen

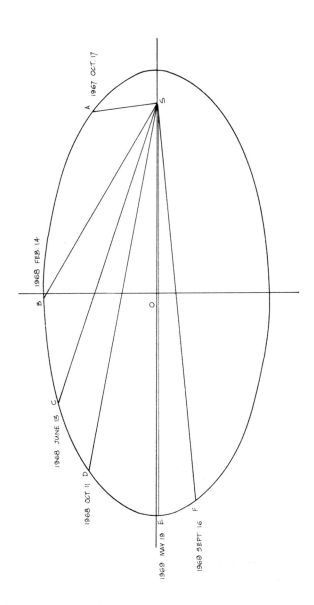

A 1967 OCT. 17

1968 FEB. 14 B

O

1968 JUNE 13 C

1968 OCT. 11 D

1969 MAY 19 E

1969 SEPT 16 F

S

Project 2   Comet P/Encke

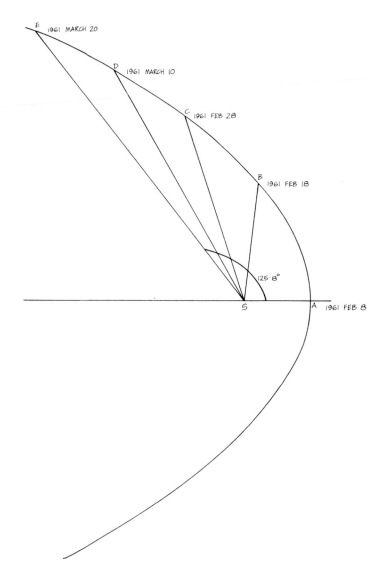

Project 3   Comet Candy

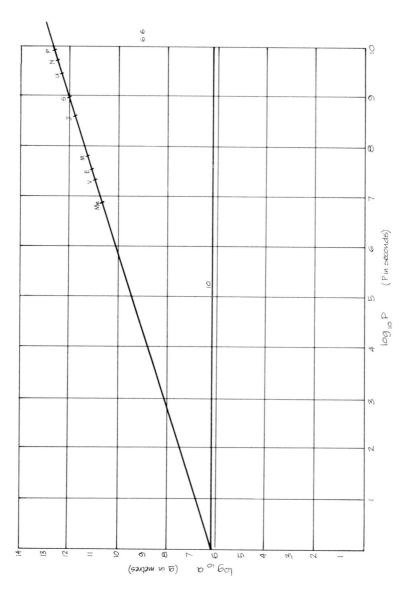

Project 4   Kepler's third law

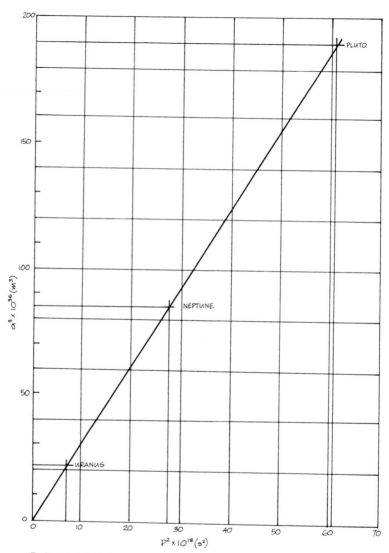

Project 5   Kepler's third law

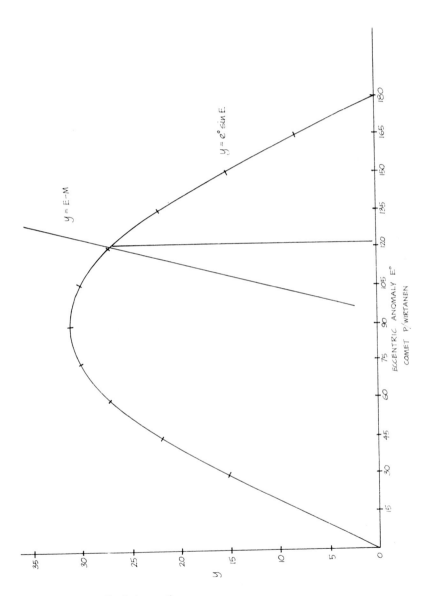

Project 6   Kepler's equation

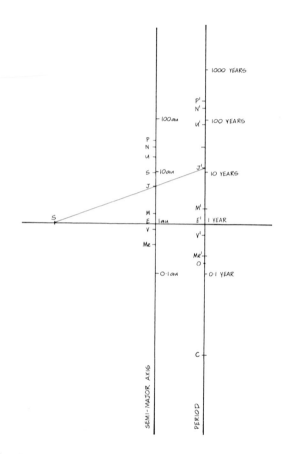

Project 7   Nomograph relating semi-major axes and orbital periods of the planets

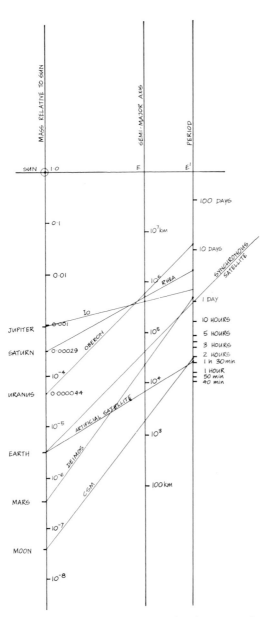

Project 8   Nomograph relating masses, semi-major axes and orbital periods of planets and satellites

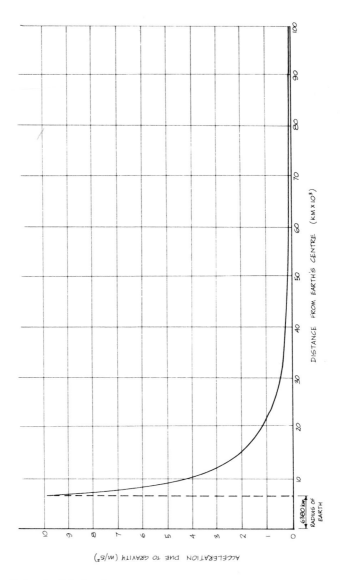

Project 9   Velocity of escape

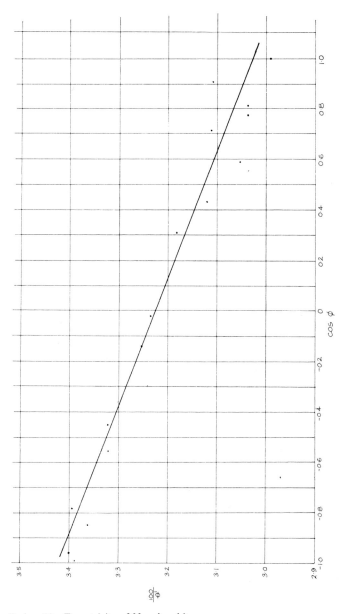

Project 10   Eccentricity of Moon's orbit

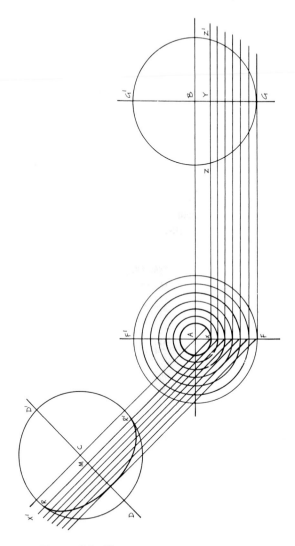

Project 11   Phases of the Moon

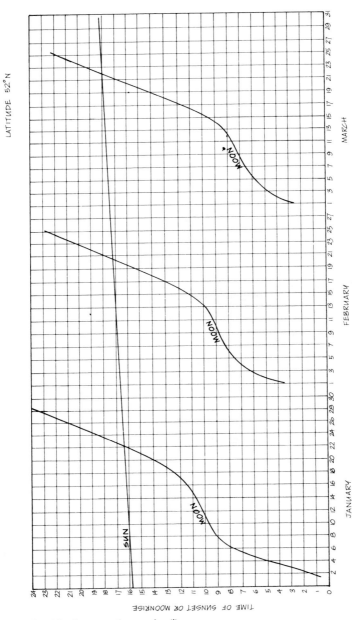

Project 13   Sunset and moonrise (i)

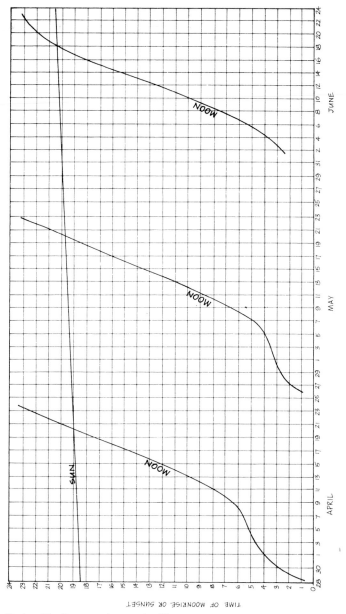

Project 13   Sunset and moonrise (ii)

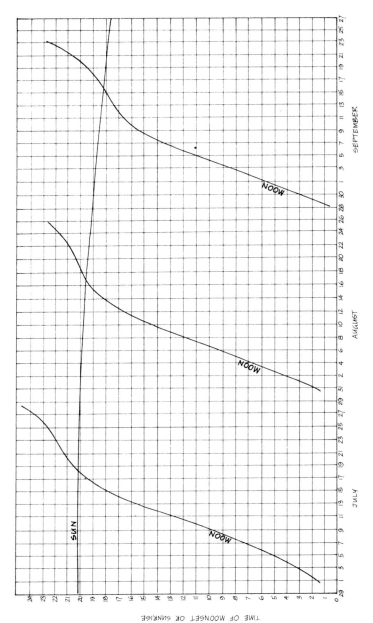

Project 13   Harvest Moon

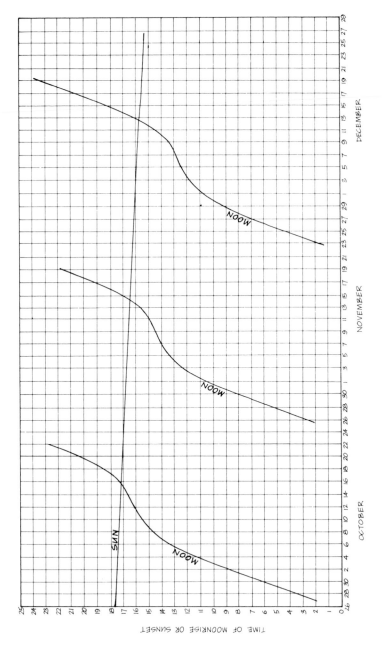

Project 13   Hunter's Moon

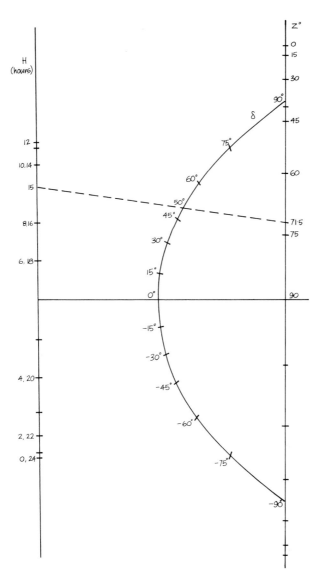

Project 14   Co-ordinate conversion nomograph (1)

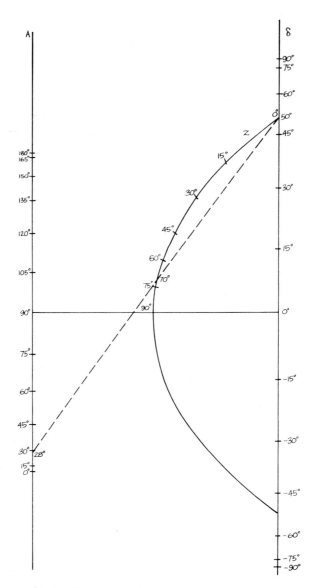

Project 14   Co-ordinate conversion nomograph (2)

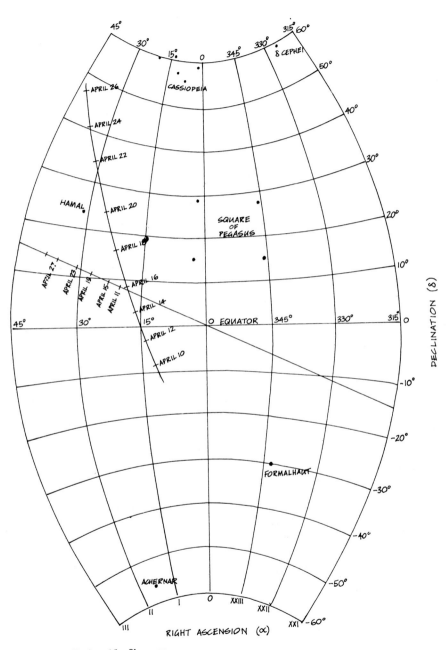

Project 15 Sky map

Project 16 Path of comet

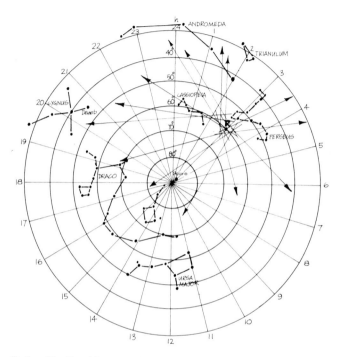

Project 18   Perseid meteor shower

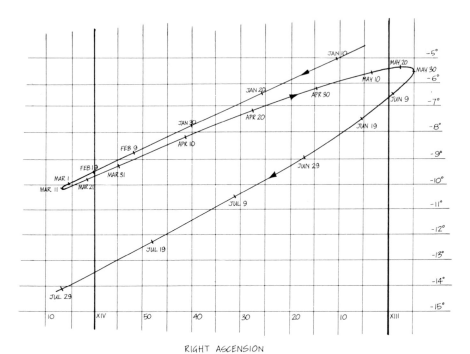

RIGHT ASCENSION

Project 19   Retrograde motion of Mars

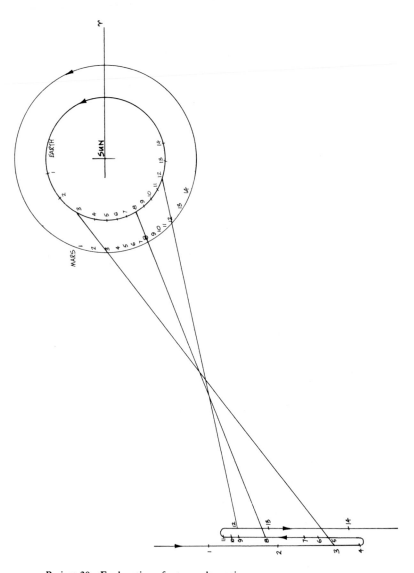

Project 20   Explanation of retrograde motion

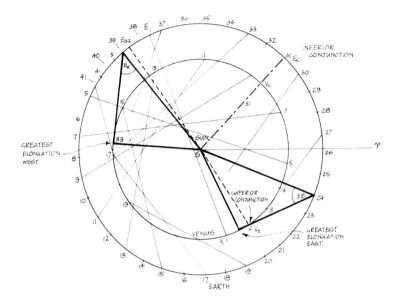

Project 21   Conjunctions, elongations and motion of Venus

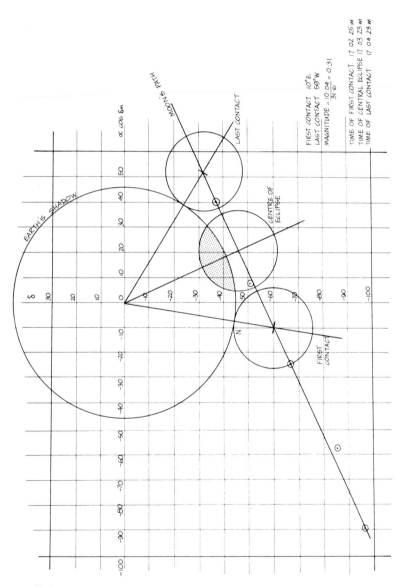

Project 23   Eclipse of the Moon

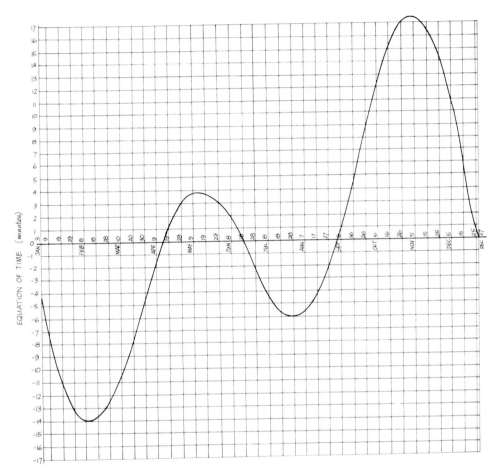

Project 24   Variation of equation of time

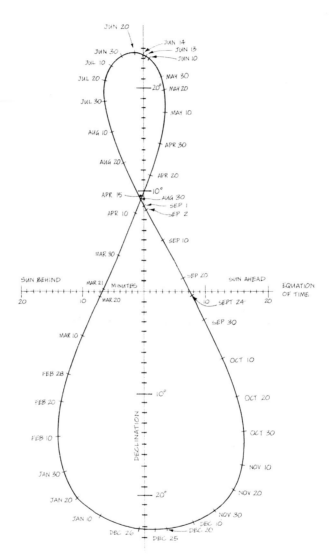

Project 25   Analemma for 1969

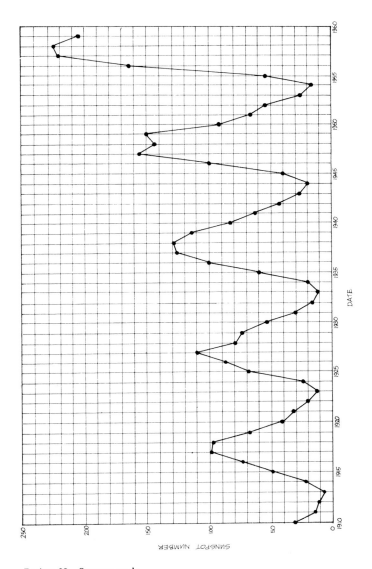

Project 29   Sunspot cycle

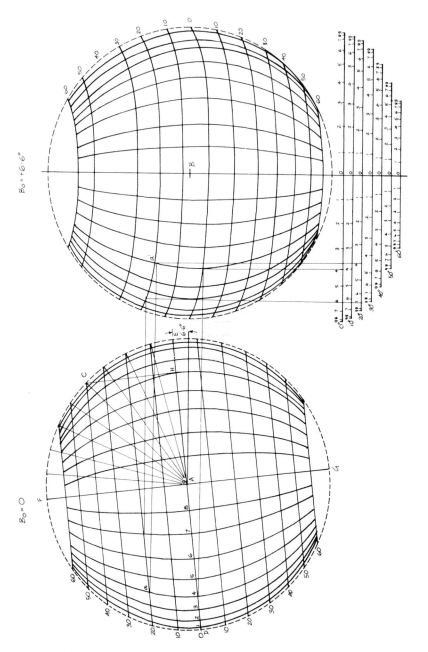

Project 30   Stonyhurst discs

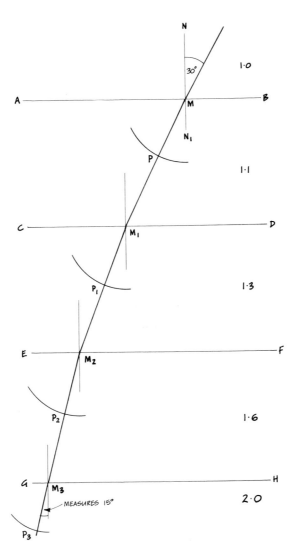

Project 31   Refraction due to layers of atmosphere

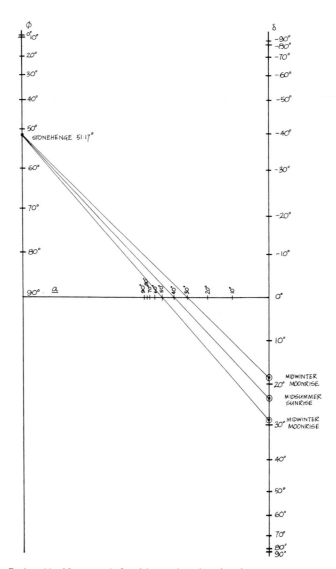

Project 32   Nomograph for rising and setting situations

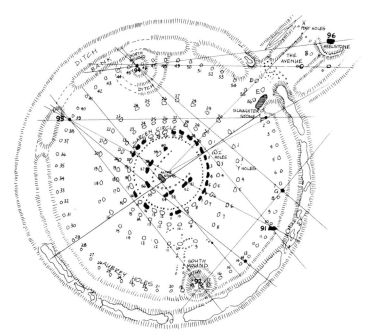

Project 33   Stonehenge alignments

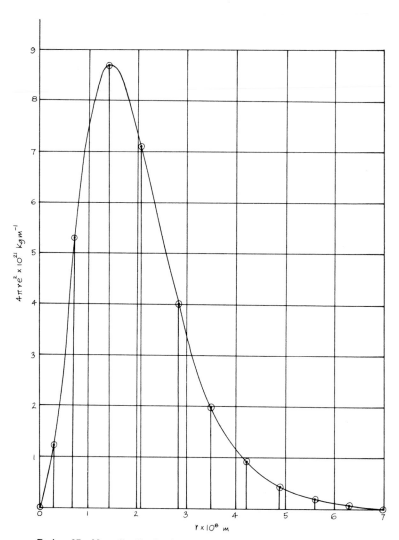

Project 37   Mass distribution in a star

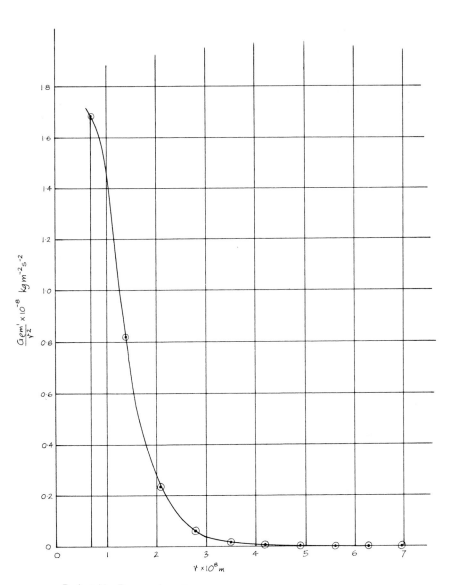

Project 38   Pressure in a star

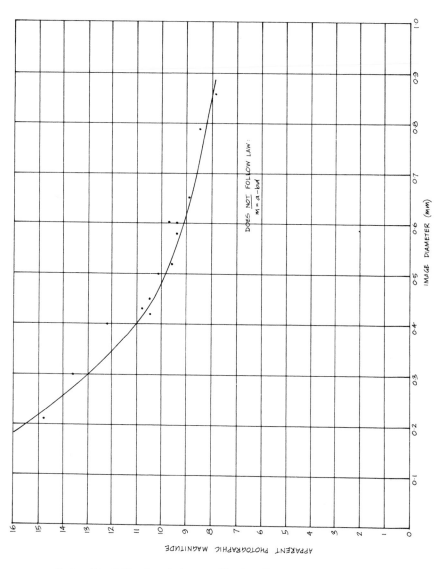

Project 39   Calibration of photographic plate (i)

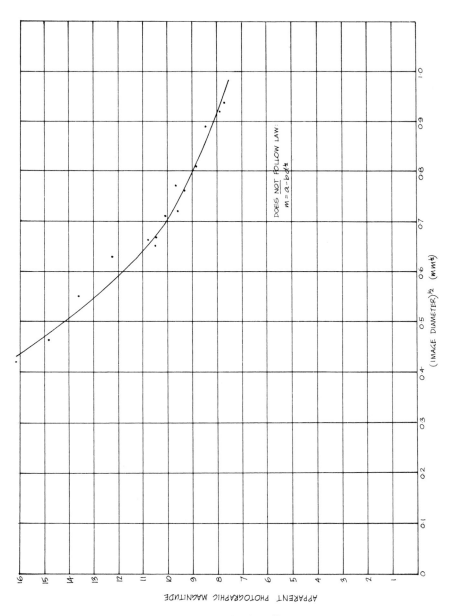

Project 39   Calibration of photographic plate (ii)

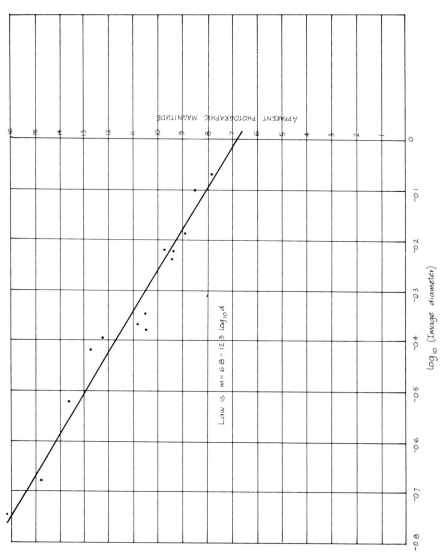

Project 39   Calibration of photographic plate (iii)

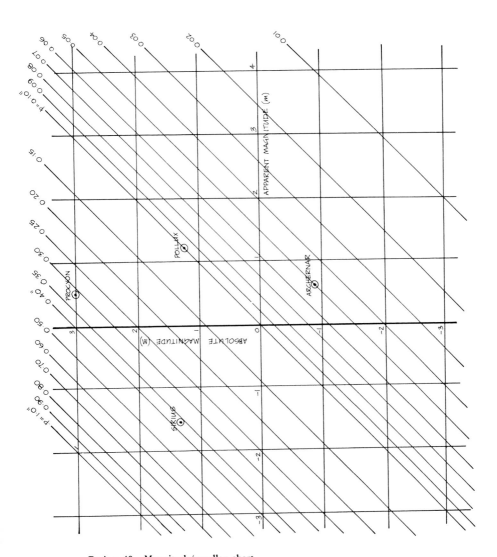

Project 40 Magnitude/parallax chart

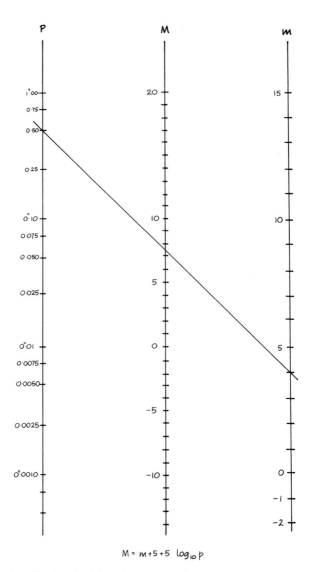

$$M = m + 5 + 5 \log_{10} p$$

Project 41   Magnitude/parallax nomograph

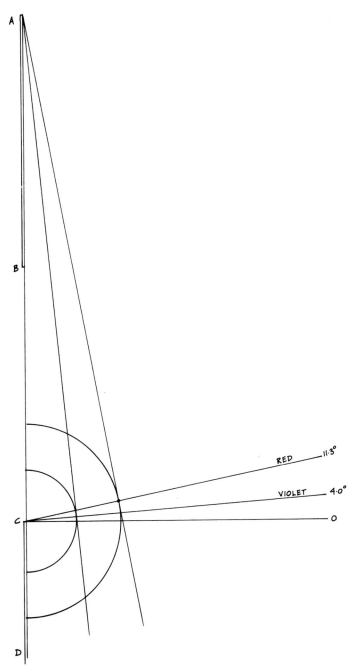

Project 42   Formation of first order spectrum (i)

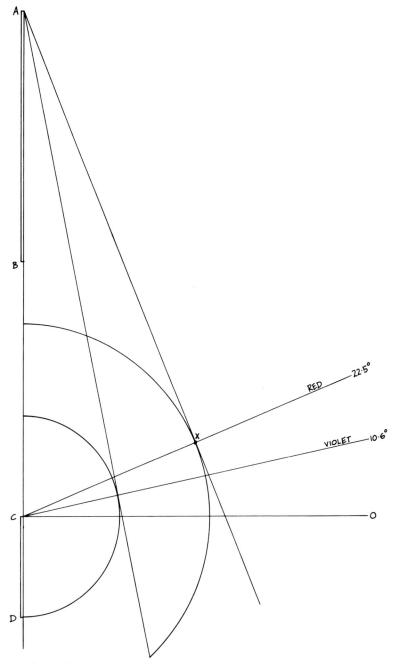

Project 42   Formation of second order spectrum (ii)

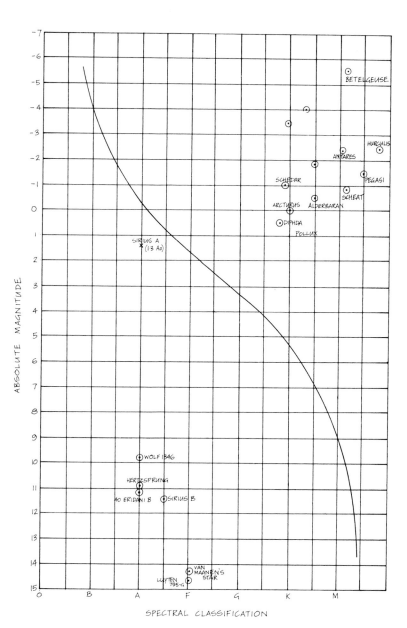

Project 45   Hertzsprung-Russell diagram

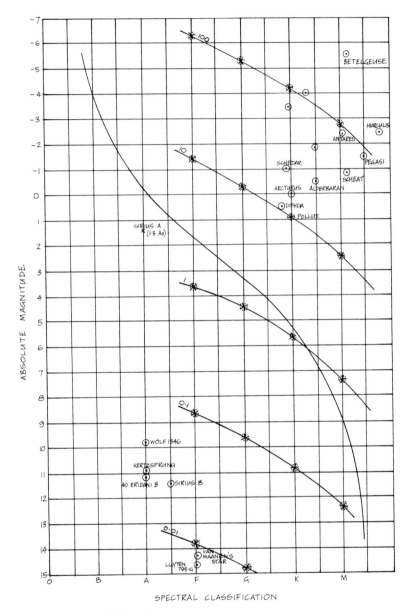

Project 46   Radius of a star

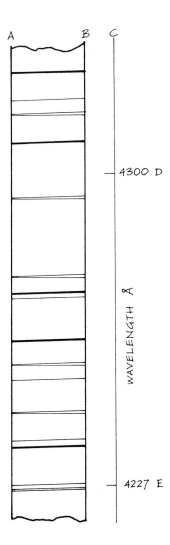

Project 47   Jupiter spectrum

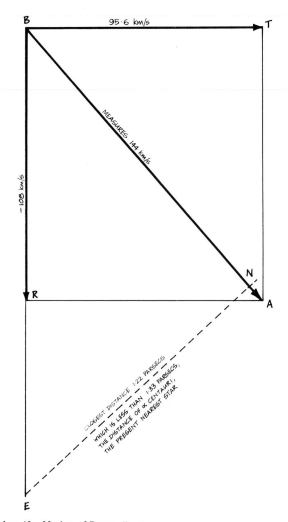

Project 48   Motion of Barnard's star

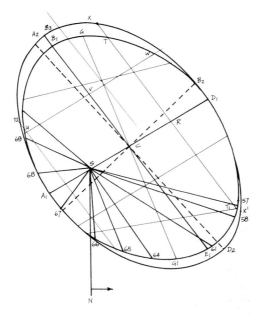

Project 49   Binary star orbit

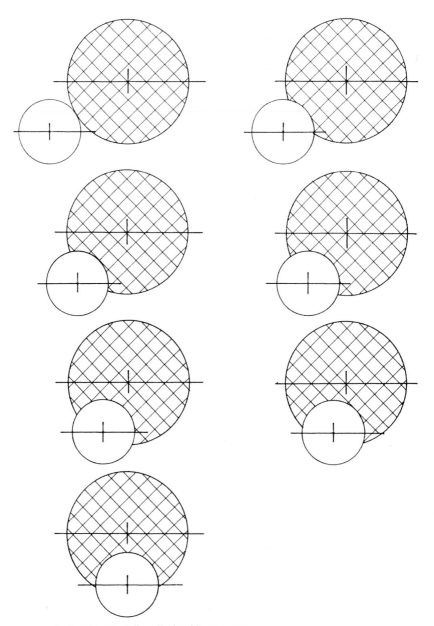

Project 50   Partially eclipsing binary system

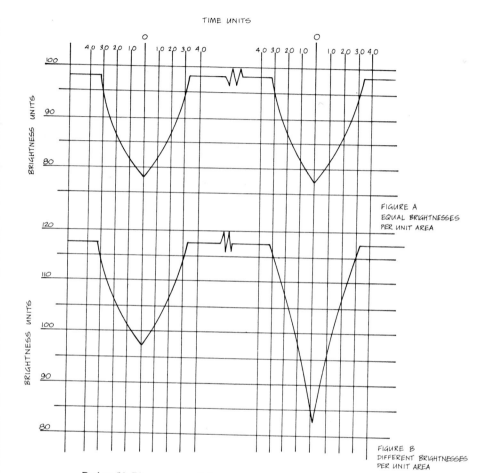

Project 50 Binary system light curves (i)

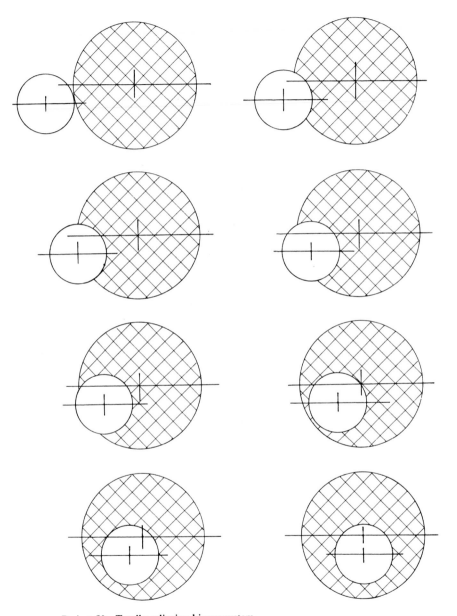

Project 50   Totally eclipsing binary system

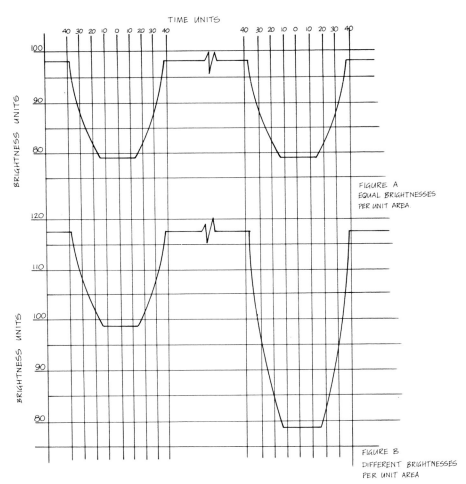

Project 50   Binary system light curves (ii)

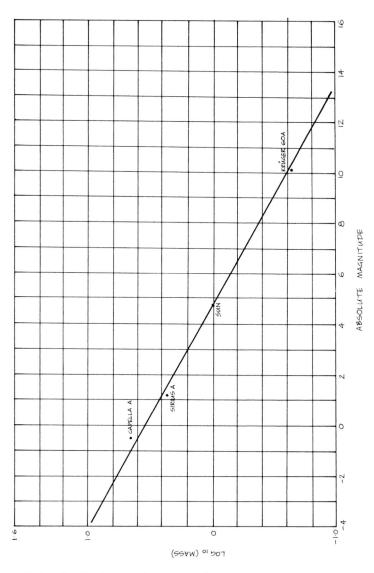

Project 51   Absolute magnitude/mass relationship

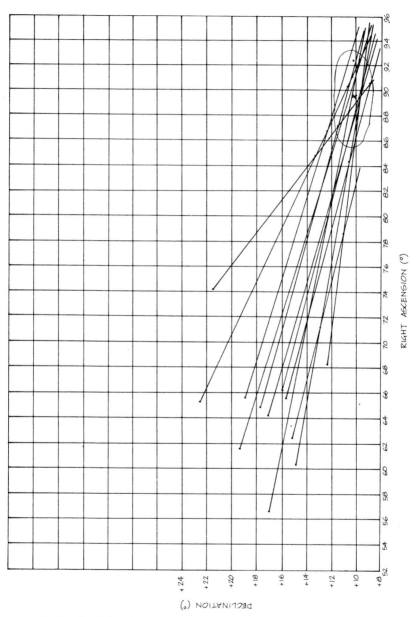

RIGHT ASCENSION (°)

DECLINATION (°)

Project 52 Point of convergence of Hyades cluster

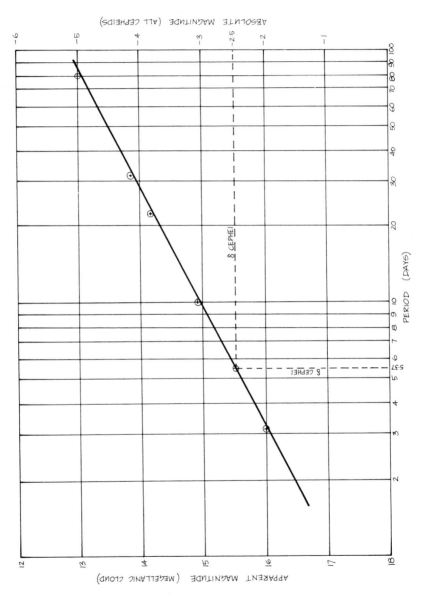

Project 53   Cepheid magnitude/period relationship

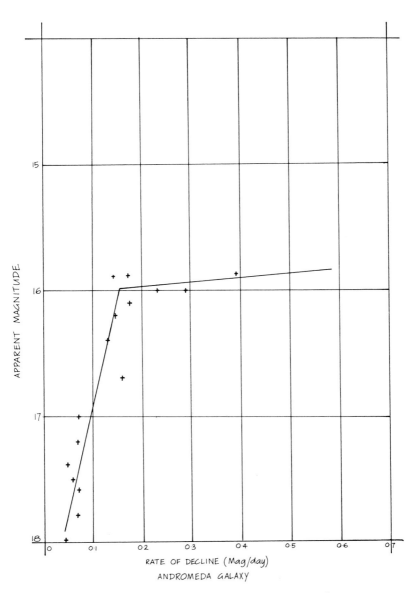

Project 54   Rate of decline of brightness of novae in Andromeda

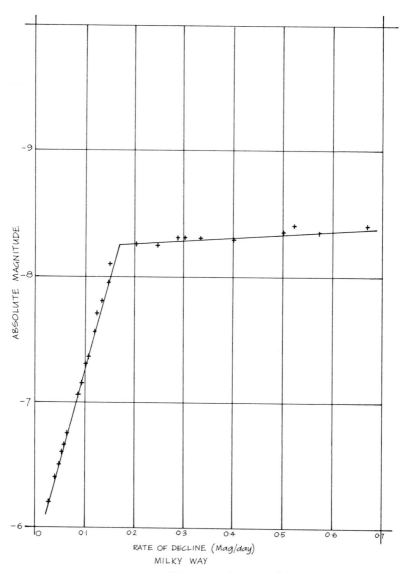

Project 54   Rate of decline of brightness of novae in Galaxy

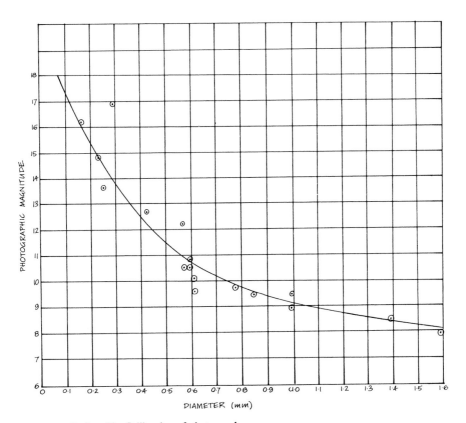

Project 55   Calibration of photograph

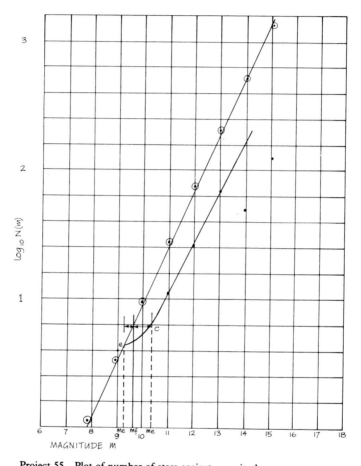

Project 55   Plot of number of stars against magnitude

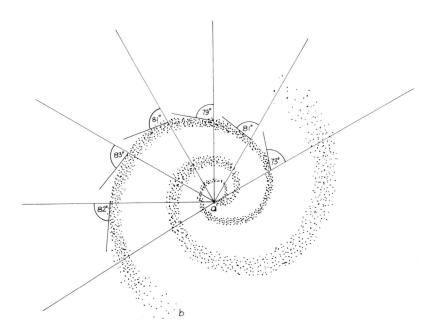

Project 56 M74 logarithmic spiral galaxy

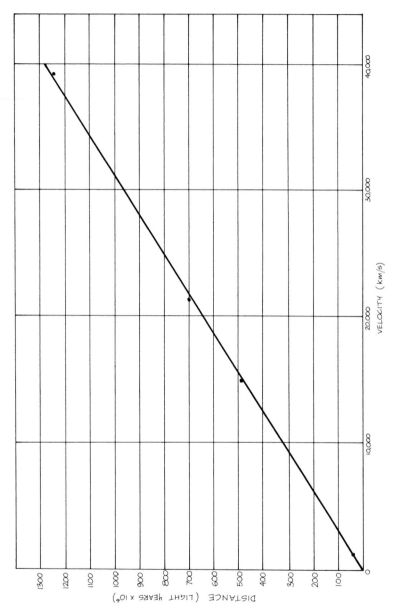

Project 57   Hubble's law

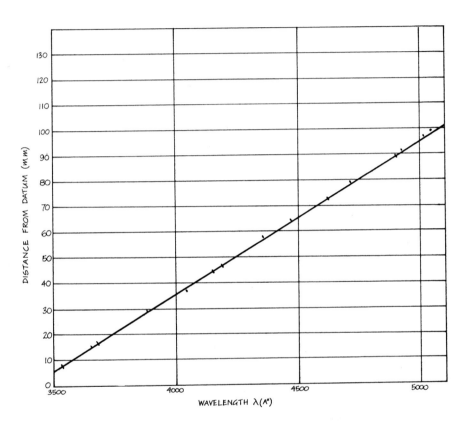

Project 58   Calibration for quasar spectrum

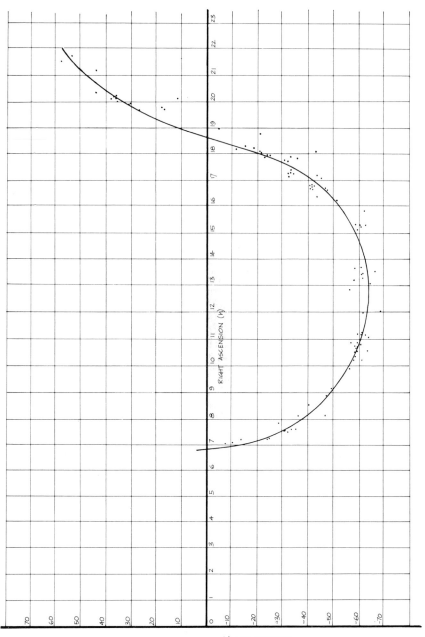

Project 59   Galactic equator

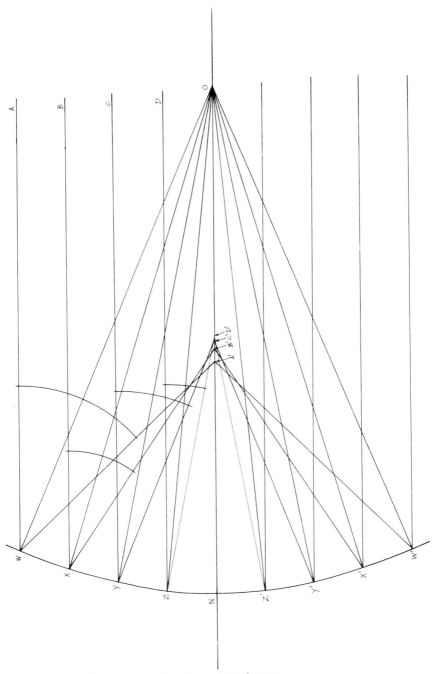

Project 60   Spherical abberation and caustic curve

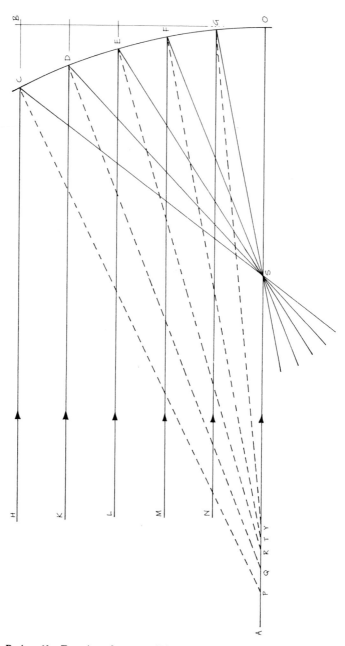

Project 61   Focusing of rays parallel to axis

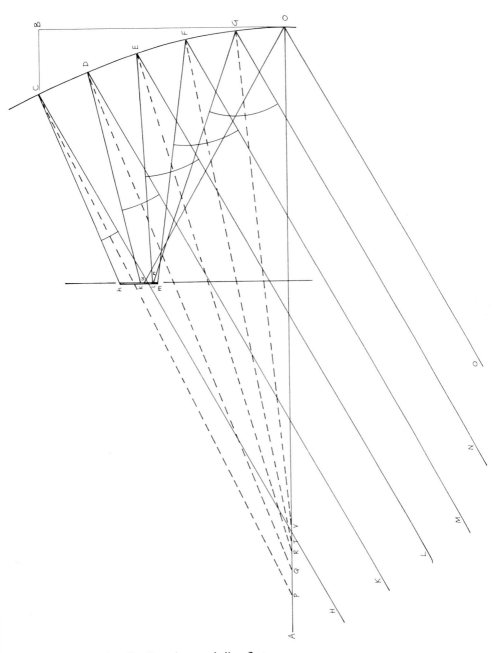

Project 61   Coma in a parabolic reflector

Index